FINANCIALLY FOCUSED QUALITY

THOMAS M. CAPPELS

S_L^t

St. Lucie Press
Boca Raton   London   New York   Washington, D.C.

Library of Congress Cataloging-in-Publication Data

Cappels, Thomas M., 1953–
    Financially focused quality / by Thomas M. Cappels.
        p.   cm.
    Includes bibliographical references and index.
    ISBN 1-57444-248-1 (alk. paper)
    1. Managerial economics.   2. Quality control.   I. Title.
HD30.22.C358   1999
658.4′013—dc21   99-28173
                                                                                   CIP

No claim to original U.S. Government works
International Standard Book Number 1-57444-248-1
Library of Congress Card Number 99-28173
Printed in the United States of America 1 2 3 4 5 6 7 8 9 0
Printed on acid-free paper

Foreword

In today's highly competitive global economy, bottom-line financial performance has become more important than ever. Rightsizing and technology improvements have created efficiencies and productivity enhancements that have led to lower costs and improved earnings. This rapid earnings growth has created expectations of unprecedented proportion on the part of shareholders, investors, and lenders. When a company fails to meet Wall Street's earnings expectations, even by a few cents, the stock is often savagely punished by the market. In today's world, understanding the bottom line and the importance of increasing shareholder value is not just a requirement for the finance organization; it is a prerequisite for all employees and disciplines within a company.

At the same time, the imperative for continued quality improvement in the products that a company provides must remain at the forefront of that company's objectives. After all, producing high-quality products that provide customers the value they are seeking is the key to long-term business success and a major determinant in beating the competition. The challenge that companies face today is how to focus their business on producing quality products, while at the same time achieving the bottom-line performance that shareholders and the investment community demand. It is exactly this challenge that author Tom Cappels tackles head-on in this comprehensive and enlightening book.

As a quality manager at Lockheed Martin, Cappels has experienced the evolution of quality first hand. He's lived the life of statistics, circles, suggestion systems, awards, MRP2, Total Quality Management, and Executive Management proclamations. He stays in tune as an author, speaker, and former officer of the American Society for Quality. In his financial roles, he has seen how an analytical, prudent, and informed financial viewpoint can reap tremendous benefits. Tom Cappels accentuates his years of hands-on quality and financial experience as a graduate-level quality management and finance professor and is president of the International CA-VIA. He has built his reputation through 20+ years of studies and publications, which culminate here in *Financially Focused Quality*.

With *Financially Focused Quality*, Tom Cappels successfully couples formal and professional education with the hard knocks of the real business world. The approaches he presents result from benchmarking the best practices of hundreds of companies. Hundreds of millions of dollars have been saved, and billions more will be, as the worlds of quality and finance unite.

I believe that *Financially Focused Quality* is an excellent book that effectively deals with one of the greatest challenges that businesses face today. It will help all members of the working class, from board members to first-line employees alike, to better understand the relationship and balance between improving quality and meeting demanding bottom-line challenges. Our futures depend on it.

John H. Schultz
Chief Financial Officer
Lindsay-Ferrari

Preface

There was a time when quality was the exclusive territory of manufacturing and operations enterprises. Today, however, quality embraces functions as diverse as engineering, marketing, purchasing, and human resources. In fact, there is not one area of business operations to which quality concepts cannot be applied. A management science has been developed to enable effective application of today's and tomorrow's quality concepts to any business for improved productivity and profitability. That science is called Financially Focused Quality (FFQ).

This text formally introduces FFQ to the business world. The FFQ management science has been evolving over the years at companies such as Lockheed Martin Missiles & Space and Litton Applied Technology and has been documented as saving these companies millions of dollars. FFQ takes Total Quality Management (TQM) and Full-Cycle Corrective Action one critical step further by ensuring an education that offers the inclusion of a financial viewpoint at the beginning of process improvement activity. This allows the integration of cost recognition in each step of the process. This book presents methodologies for company-wide and scholastic education on key financial considerations. Quality management and quality principles are reviewed. Basic financial concepts are explained, then FFQ elements and their use in making and implementing business decisions are introduced, and, finally, case studies show successful FFQ applications. Readers will be equipped to implement FFQ in their work environments, offering the prospect of improving the translation of quality and cost-cutting measures to the bottom line.

Financially Focused Quality provides management and quality professionals with a financial foundation and basic business concepts that are vital

The Author

Thomas M. Cappels is an Adjunct Professor for the University of Phoenix, developing and teaching graduate level business and quality management courses. He is also responsible for reporting savings for the Lockheed Martin Missiles & Space (LMMS) Lean Process Center (LPC), the goal of which is to reduce operating costs substantially. Cappels has held many responsible positions in contract administration and finance. He has been responsible for negotiation and administration of management and information resources supporting such high-profile Lockheed Martin programs as the Hubble Space Telescope, the International Space Station, IRIDIUM, and MILSTAR. He joined the American Society for Quality in 1979 and serves as president of the 400,000 member Computer Associates Visual Information Association.

Cappels received his Bachelor's degree in Communications and his Master's degree in Business Administration from San Jose State University. Cappels earned his California State Teaching Credential and Second Class Federal Communications Commission License in the 1970s and has taught courses in business, communications, quality, and finance for over 25 years. He has presented many papers, appeared on radio and television, lectured at numerous educational functions, and authored over 100 publications, including contributions to *Quality Progress Magazine*. His book, *Full-Cycle Corrective Action: Managing for Quality and Profits*, is published and distributed internationally by the ASQ Quality Press.

Contents

Chapter 1. Update on Quality 1
Introduction 1
Focus on Quality 2
Quality Becomes Big Business 3
The Business of Quality 4
Evolution of the Corporate World 5
Financially Focused Quality 6
Self-Study/Discussion Questions 6
References 7

Chapter 2. Employee Involvement in Quality Management 9
Introduction 9
Human Resources 10
    Employee Benefits 10
        Retirement Plan 10
        Medical and Dental Benefits 11
        Vacation/Holiday Benefits 11
        Sick Pay/Salary Continuation Policy 11
        Employee Stock Ownership Plans 11
        Employee Stock Options 12
        Assistance With Further Education 12
        Company Training Programs 13
        Flexibility of Shifts 13
        Transportation and Commute Alternatives 13
        Employee Recreation Facilities 13
        Executive Perks 14
    Summary 14
Employee Participation in Quality Management Activities 14
    Total Quality Management 15
    Total Quality Service 15
    Empowerment and Ownership 16
    Employee Suggestion Systems 17
    Cost Reduction or Avoidance Programs 18

Continuous Quality Improvement 18
Process Management 19
Pay for Performance 19
Participative Management and Team Building 20
Quality Circles 20
Task Force Teams 21
Natural Work Teams 21
Process Improvement Teams 22
Effective Meetings 22
Meet at the Right Time and Place 25
Meeting Locations 26
Possible Roles of Attendees 26
Brainstorming 27
Affinity Diagrams 28
Lean Process 28
Kaizen Event 28
Six Sigma 28
Summary 29
Self-Study/Discussion Questions 29
References 29

Chapter 3. Managerial Strategies for Quality Management 31
Introduction 31
Total Improvement Management 32
The Financially Focused Quality Blueprint 32
Component 1: Direction 33
Component 2: Basic Concepts 34
Component 3: Profit Generation 34
Component 4: Enterprise Impacts 35
Component 5: Rewards and Recognition 35
Manufacturing Resource Planning 35
Just-in-Time Manufacturing 36
Business Objectives 36
Vision Statements 36
Mission Statement 39
Benchmarking/Best Practices 39
Training 40
Measurement Process 41
Core Competencies 42
Outsourcing 42
Reengineering 43
Restructuring 43
Rightsizing/Smartsizing 44
Downsizing and Budget Cutting 44
Partnerships With Suppliers 45
Quality Costs and Poor Quality Costs 45
Full-Cycle Corrective Action 46
Cost as an Independent Variable 46

Summary 46
Self-Study/Discussion Questions 46
References 47

Chapter 4. Quality Management Systems and Customer Focus 49
Introduction 49
Quality Management Systems 50
Standardization 50
    The Euro 50
    Movie, Video, and Computer Products 51
        8-mm Movies: Super 8 vs. Single 8 Movie Formats 51
        Beta vs. VHS Videotape Formats 51
        VHS vs. 8-mm Videotape 51
        Macintosh vs. Personal Computers 52
ISO 9000 and ISO 14000 Series 52
    ISO 9000 Series 53
    ISO 14000 Series 54
External Customer Focus 55
Customer Feedback 56
    Expensive Products and Customer Feedback 56
    Inexpensive Products and Customer Feedback 57
    Service Industry Feedback — Surveys and Questionnaires 58
    Service Industry Feedback — Real Time (Restaurant) 58
    Service Industry Feedback — Real Time (Hotel) 60
    Unsolicited Letters 60
Summary 61
Self-Study/Discussion Questions 62
References 62

Chapter 5. Quality Control 63
Introduction 63
Dr. W. Edwards Deming 63
Process Quality Control and Statistical Analysis 65
Diminishing Returns of Quality Control Measures 65
Quality Control Definition and Goals 67
Quality Assurance Organization 67
Closed-Loop Corrective Action 68
Relationship of Financially Focused Quality to Closed-Loop Corrective Action 70
    CLCA Operations and Failure Reporting Systems 70
    Supplier Operations 70
    Fleet and Field Facilities Operations 74
    Quality Engineering 74
    Summary 76
Quality Assurance and the Product Life Cycle 76
    Research and Development Phase 76
    Design Phase 76
    Production Phase 77
    Operation Phase 77
    Disposal Phase 77

Cycle of Quality Activities 77
Quality Assurance Functions 78
    Receiving Inspection 78
    In-Process Inspection 79
    Shipping Inspection and/or Final Inspection 79
    Accelerated Test 79
    Configuration Management 79
    Destructive Testing 80
    Engineering Evaluation Test 80
    Electrical In-Process Inspection 80
    Fabrication and Assembly Inspection and Test 80
    Failure Analysis 80
    Failure Diagnosis 80
    Failure Verification 81
    First Article Compatibility Test 81
    Material Review Board 81
    Mechanical In-Process Inspection 81
    Metrology Audits 81
    Parts Application Review 81
    Product Evaluation Test 81
    Quality Assessment 82
    Quality Audit 82
    Receiving Inspection and Test 82
    Source Acceptance 82
    Source Verification Inspection 82
    Statistical Process Control 82
    Value Engineering 82
    Summary 83
Traditional Finance Involvement in the Cycle of Quality Activities 83
Completing the Cycle of Quality Activities: Finance 85
Self-Study/Discussion Questions 86
References 86

Chapter 6. Avoiding Product Liability: Design and Development 87
Introduction 87
Product Safety 87
Responsibility for Product Safety 88
Manufacturer Liability 88
Basic Conditions of a Product Liability Case 89
Manufacturing Defects 90
Quality Assurance vs. Product Liability 90
Evolution of Quality Assurance 91
Relationship Between Quality Assurance and Product Liability 92
Systematic Approach to Quality and Safety Assurance 92
Product Design Safety 93
Purchased Material Quality 93
Manufactured Product Quality 93
Product Safety After Delivery 94

Quality Control and Product Liability Aversion 95
Product Safety-Related Quality Information 95
The Financially Focused Approach to Product Safety-Related Corrective Action 96
Product Design: The Beginning 96
    Product Design Safety 96
    Product Safety Specification 97
    Standards for Design and Manufacture 97
    Evaluation of Safe Product Design Quality 98
    Design and Documentation Reviews 98
    Safety Design Review 99
Classification of Safety Critical Characteristics 100
Design and Qualification Test 100
Vendor Materials Control 101
    Sources for Raw Materials, Parts, and Components 101
    Make-or-Buy Committees 101
    Supplier Selection 102
    Ordering Cost-Effective Quantities 102
    Conforming to Requirements 102
    Suppliers' Performance Histories 103
    Supplier Quality Evaluation Survey 103
    Supplier Evaluation Questionnaire 104
    Purchasing Provisions and Specifications 104
Summary 105
Self-Study/Discussion Questions 105
References 106

Chapter 7. Avoiding Product Liability: Manufacturing and Use 107
Introduction 107
Inspection and Test of Vendor Components 107
    First Article Inspection 108
    Source Inspection 108
    Vendor Inspection 109
    Receiving Inspection 109
Inspection Records During the Receiving Cycle 110
Source/Receiving Inspection: Corrective Action 110
Inspection of Storage Facilities 110
Inspection and Testing in Manufacturing 111
The Obligation To Inspect and Test 111
Inspection and Test Areas 111
Measurement 112
Training of Inspectors 113
Inspection Instructions 114
Inspection Procedure Manuals 115
Industrial Quality Assurance 116
Assessing the Production Process 116
    Process Capability Studies 116
    Mandatory Process Evaluation 116
The Manufacturing Process 117

Process Control 117
In-Process Inspection and Testing 118
Tolerances 119
Nonconforming Material 119
Quality Assurance Through Statistics and Sampling 120
    Definition of Terms 120
    Sampling Inspection 121
    Random Numbers 121
    Sampling Plans 123
    Degree of Inspection 124
    Sampling Plans vs. 100% Inspection 124
    Arguments Against Sampling Plans 125
Inspector Errors 125
Consumerism 126
    Shipping Inspection 126
    Packaging for Product Safety 126
Product Safety in the Field 127
    Service Publications 127
    Field Results 127
Product Recall 128
Management Resources for Assuring Quality 129
    Accredited Laboratories 129
Audit Control 130
    Vendor Audit 130
    Product Safety Audit 131
Quality Assurance Evaluation Checklist 132
Summary 133
Self-Study/Discussion Questions 134
References 134

Chapter 8. Financial Administration and
    Financially Focused Quality Training 137
Introduction to Financially Focused Quality 137
Financial Training 138
Cycle of Financial Activities 139
Financial Concepts 141
    Accounting Classifications 142
    Direct vs. Indirect Costs 142
    Labor vs. Nonlabor Costs 143
    Direct-Direct and Allocated-Direct Costs 144
    Overhead Costs (Indirect Costs) 144
    Work Breakdown Structure 144
    Cost Accumulation Structure 145
    Charge Numbers 145
Timekeeping, Timecards, and Cost Segregation 146
    Virtual and Manual Timecards and Labor Recording 148
Forecasting Personnel 150
    Criteria 150

Proposals 150
Training 150
Facilities 150
Human Resources 151
Forward Pricing Rates/Overhead and General & Administrative Costs 151
Forecasting Direct Personnel 151
Yield/Realization Factor 152
Forecasting Problem 154
Contracting Modes (Fixed Price/Cost Reimbursable) 154
Fixed Price 154
Firm Fixed Price 154
Fixed Price Incentive 155
Cost Reimbursable Contract 155
Cost Plus Fixed Fee 155
Cost Plus Incentive Fee 155
Cost Plus Award Fee 156
Disallowances 156
Rates per Direct-Labor Hour 156
Teaching Financially Focused Quality and Financial Concepts 157
An Effective FFQ Training Guideline 158
Examples of Financially Focused Training Materials 160
Self-Paced Training 160
Summary 160
Self-Study/Discussion Questions 161
References 161

Chapter 9. Financial Functions 163
Introduction 163
Proposals and Pricing 163
Commercial Contracting 163
Government Contracting 164
Proposal Preparation Flow 165
Raw Resource 166
Proposal Components 166
Basis of Estimate 166
Elements of a Historical Basis of Estimate 168
Engineering Estimates 168
Security Guidelines for Cost Proposals 168
Audit Checklist for Basis of Estimate 169
Pricing 170
Budgeting 170
Government Appropriations and Funding Fences 171
Management Reserve and Undistributed Budget 171
Reporting 172
General Accounting 173
Payroll Accounting 173
Accounts Payable 174
Financial Accounting 175

Cost Accounting 175
Contract Closure 176
Summary 176
Self-Study/Discussion Questions 177

Chapter 10. Financially Focused Quality Performers and Components 179
Introduction 179
Financially Focused Quality Overview 180
    Failure Identifier or Opportunity Identifier 181
    Failure Notice 181
    Process Improvement Recommendation 181
    Process Improvement Coordinator 181
    Failure Analyst or Process Analyst 181
    Process Analysis Meeting 182
    Finance 182
    Process Improvement Cost Analysis 182
    Process Improvement 182
    Process Improvement Follow-Up Plan 182
    Product Support 183
    Process Improvement Closure Notice 183
Failure or Opportunity Identifier 183
    Process Improvement Begins With Opportunity Identification 185
    Methods for Failure and Opportunity Identification 185
The Failure and Opportunity Identification Process 186
    Customers 188
    Government Failure Identification 188
    Warranty Service Department 189
    Failure Identifiers in the Factory 189
        Inspectors 189
        Manufacturing Employees 189
        All Other Company Employees 189
    Summary 190
Failure Notice 190
    Failure During Manufacturing 190
    Failure: Customer Return 191
    Failure: Hotel Industry 192
    Generic Failure 192
Process Improvement Recommendation 193
    Process Improvement Recommendation Success Factors 195
        Ease of Processing 195
        Low Implementation/Administration Costs 195
        Mandatory Management Involvement 196
    Process Improvement Recommendation Motivation 198
    Examples of Process Improvement Recommendations 200
    Illustration of Acting on a Process Improvement Recommendation 201
Summary 202
Self-Study/Discussion Questions 202

**Chapter 11. Financially Focused Quality Process
Improvement Coordination and Decision-Making** 203

Introduction 203
Process Improvement Coordination 203
    When Process Improvement or Corrective Action Is Not Required 204
    Time Requirements for Process Improvement Coordination 205
    Time Requirements for Process Improvement
    Recommendation Coordination 205
    Time Requirements for Failure Notice Coordination 205
    Process Improvement Recommendations: Related Functions
    of the Process Improvement Coordinator 206
The Financially Focused Quality Mindset 206
    The Process Improvement Coordinator and the Failure Notice 207
        Auto or Aircraft Manufacturing Failure 207
        Television Manufacturing 208
        Service Industry Failure #1 208
        Service Industry Failure #2 208
The Coordination Process 208
    Coordination of Process Improvement Recommendation Projects 209
    Coordination of Process Improvement Recommendation Concerns 210
    Coordination of Failure Notices 212
Failure Analysts 212
Process Analysis Meeting 214
Brainstorming 216
    Topic or Objective Definition 217
    Recording Ideas 218
    Presentation of Rules 218
        Take Turns 218
        Place Maximum Effort on Idea Generation 218
        Don't Throw Anyone into a State of Self-Conscious Distress 218
        Everyone Has To Participate 219
        Relax and Have a Sense of Humor 219
        Exaggerate 219
        Condense Ideas and Be Concise 219
        No Judgments Allowed 219
    Completion of Brainstorming and Discussion 220
Financially Focused Quality and Probabilities 220
    Selection 221
    Components of Effective Brainstorming 221
Process Improvement Cost Analysis 222
Process Improvement Follow-Up Plan 224
Process Improvement Closure Notice 226
Financial Administration 227
Product Support Organizations 229
Summary 230
Self-Study/Discussion Questions 231
References 231

Case Study A: Perky Pets — Commercial Manufacturing 233

Case Study B: Financially Focused Quality
    Implemented in Software Engineering 245

Case Study C: Bob's Purple Bayou Café 257

Case Study D: One Company's Success With Outsourcing 265

Case Study E: Hotel Operations 273

Index 287

1 Update on Quality

Introduction

You're a contestant playing the television game show *Jeopardy* and being videotaped before a live studio audience. The category is "Those Darned Quality Acronyms," the amount is $1000, and the answer is

It was the dominant approach to Quality Management in the 1990s.

Trying to ring in *before* Alex Trebek has finished reading the answer causes frustration, because you keep clicking the button and nothing happens. Alex finishes reading the answer, and the opponent to your right — defending champion Chuck Forrest — rings in with the question with perfect timing:

"What is TQM?" Chuck declares confidently.

"That's right," Alex states. "Total Quality Management. Pick again!" But, you disagree and your frustration overcomes you. "NO!" you scream. "It's *not* TQM. It's FFQ — Financially Focused Quality!" Without skipping a beat, Alex says, "We'll check with our judges during this short commercial break. Don't go away!"

It is not surprising that the *Jeopardy* judges will have to ponder this issue. The science of quality has experienced a dramatic revolution during the past 20 years. In the 1980s and 1990s, the economic, political, and environmental spectra experienced a tremendous awakening in regard to the importance of "quality". Once perceived as merely the absence of failures, "quality" became widely credited with playing a critical role in *all* aspects of business. This led to wide recognition and acceptance of such management strategies as "Reengineering", "Continuous Improvement", and "Total Quality Management".

Focus on Quality

Enhanced quality consciousness increased the demand for seminars, publications, specialized training, and consultants. Millions of dollars were spent annually on programs to improve quality and productivity. Below is a partial listing of approaches utilized by companies in the 1980s and '90s to increase profitability:

- Continuous Improvement
- Total Quality Management
- Total Improvement Management
- Benchmarking
- Best Practices
- Customer Focus (external and internal)
- Customer Perception
- Customer Satisfaction
- Brainstorming
- Effective Meetings
- Quality Management Systems
- ISO 9000 series
- ISO 14000 series
- Team Building
- Affinity Diagrams
- Interrelationship Diagrams
- Tree Diagrams
- Malcolm Baldrige National Quality Award Criteria
- Matrix Diagrams
- Prioritization Matrix
- Process Decision Program Charts
- *Kaizen* Events
- MRP
- MRP2

- Just-in-Time Purchasing
- Business Process Improvement
- Employee Stock Ownership Plans
- Employee Suggestion Systems
- Six Sigma
- Continuous Quality Improvement
- Empowerment
- Quality Circles
- Employee Participation
- Lean Process
- Managers as Team Builders
- Cost Reduction/Avoidance Programs
- Process Improvement
- Mission Statements
- Vision Statements
- Communications: Town Hall and Brown, newsletters (e.g., e-mail)
- Process Management
- Rewards and Recognition
- Pay for Performance
- Cost of Quality
- Poor Quality Cost
- Restructuring
- Reengineering
- Rightsizing
- Downsizing
- Outsourcing
- Full-Cycle Corrective Action

Quality Becomes Big Business

There is no doubt about it. In the 1990s, the pursuit of quality became Big Business. Thousands of consultants began offering their services for such quality programs as ISO certifications, Continuous Improvement, and Total Quality Management implementations. Expensive seminars and conferences were very well attended, as millions of dollars were spent in an effort to learn the latest quality innovations.

The Business of Quality

As quality awareness grew, many companies began to lose sight of the business implications of quality which resulted in *decreased* profitability. Many of the programs discussed in this text have been embraced internationally in the past two decades. Recently, however, it has become clear that such strategies *alone* are not the panacea that believers had been expecting.

Newsweek cited two formal surveys reflecting poor results of TQM activities. The first, conducted by Rath and Strong of Lexington, graded companies on TQM efforts to improve market share, rein in costs, and make customers happy. A review of the survey shows that most companies rated D's and F's. Boston's Arthur D. Little conducted the second survey. Here, 500 companies were examined, and only 36% said that the process was having a significant impact on their ability to quash competitors.

Some of the companies surveyed even complained that such management techniques cost more than they were worth. *Newsweek* cited the case of a company so obsessed with improving its inventory process that it spent a fortune on a state-of-the-art computer system. This major investment resulted in a paradox: the wholesale cost of producing a 25¢ item soared to a ridiculous $2.89.

Another paradox exists with the Malcolm Baldrige National Quality Award. H. James Harrington reports, for example, that one Houston, TX, organization went bankrupt just months after winning the award, while another winning organization's stock dropped 75%. That company let hundreds of thousands of employees go and cut dividends by 75% after winning.

The above reports and surveys suggest that something is missing in the quality management philosophies adopted by those that run companies. Management wants to please the owners of their companies, and, while that implies a desire for consistent quality and the latest technological advances, the number one priority for business owners (e.g., stockholders) is to *maintain and/or increase profits*. In the stock-price-driven U.S. economy of the late '90s, massive corporate layoffs no longer seem to be seen as a sign of economic slowdown. Rather, they now are perceived to be adjustments to an economy seeking to keep corporate profits up and to be more productive and globally competitive.

Evolution of the Corporate World

Today's corporate culture continues to embrace the latest TQM-like activities, and, while the quality focus remains, there has been an increased focus on the bottom line. The comfortable environment of empowerment, teamwork, and customer orientation has been changing to make room for reengineering, restructuring, downsizing, and outsourcing. Such actions will often reduce costs (exclusive of severance expenses) in the short term; however, other related impacts (e.g., substandard quality, increased training, decreased employee morale) can have significant negative impacts on profitability.

In just the two months of October and November 1997, major companies took the following actions:

1. Citicorp (CCI) said it would dismiss 9000 of its 90,000 employees worldwide as part of a massive restructuring designed to cut costs and improve the efficiency of back-office operations. The layoffs, which were to take place over the next 18 months, would be offset by the creation of 1500 new jobs, bringing total cutbacks to 8.3% of the bank's work force.
2. Silicon Graphics, Inc., the high-end graphic computer concern, announced it would be firing up to 1000 workers.
3. Fruit of the Loom gave 60-day layoff notices to nearly 4200 Louisiana workers and another 1035 in Kentucky, and announced it was closing its plant at Abbeville. This was part of a nationwide employment cutback in an effort to send more jobs to Central America, where the wage levels are more competitive.
4. Cadbury Schweppes' U.S. beverage unit, Dr. Pepper/Seven Up, Inc., announced it would cut 10% of its U.S. work force (about 110 workers) by the end of 1998.
5. Levi-Strauss & Company, the blue jeans manufacturer, has said it will shut down 11 of its 37 plants and cut 34% of its North American labor force. The layoffs at the clothing concern were expected to be 6395 people out of a global work force of 37,500.
6. Imaging giant Eastman Kodak Company said it might cut 14,000 jobs, slash costs by as much as $1 billion, consolidate several businesses, and expand joint ventures.

One year later, in December 1998, the trend was still ongoing. Deutsche Bank AG announced plans to acquire Bankers Trust Corporation for $10.1 billion, forming the world's biggest financial services company and promising to boost profits by cutting 5500 jobs, or 5.7% of staff. One week later, Citigroup (formed by the merger of Citicorp and Travelers Group) announced that it would cut another 1400 jobs, bringing the total to 10,400, or 6.5% of its work force.

The Bloomberg News Service reported that U.S. companies announced more than 574,000 job cuts in the first 11 months of 1998. Rightsizing (i.e., downsizing and layoffs) continues. Virtually every company is now looking beyond "quality" to achieve financial goals. "Restructuring", "reengineering", "outsourcing", and "rightsizing" are popular terms being used in corporate boardrooms.

Granted, every single one of the above-listed approaches can be successfully applied to increase profitability. However, the key to successful application is incorporating an educated financial focus.

Financially Focused Quality

The next step in the evolution of the ultimate quality and productivity improvement program leads to Financially Focused Quality (FFQ). Applying the tools and principles of FFQ with any of the above management approaches will minimize the risk of adverse affects on profits, and greatly enhance the opportunities for increased productivity and profitability.

Corporate executives justify costcutting measures by focusing on the cold cruel world of balance sheets and the bottom line. Headcount reductions and other actions designed to improve the business are dictated and budgets are slashed prior to reaching a thorough understanding of their total impacts to profitability. It is now paramount that management, administrative, and technical employees everywhere understand from an educated, financial perspective how their activities and decisions affect the bottom line. FFQ offers the means to achieve this understanding.

Self-Study/Discussion Questions

1. In regard to the approaches to improving quality and productivity listed in this section:

    a. Which have been applied where you work?

    b. What are the names and descriptions of other strategies implemented?

    c. What sort of investments in time and money did management expend?

    d. Did the approach work? In your mind, was it worth the investment?

    e. Contrast the different techniques. What are advantages and disadvantages of each?

2. In regard to mergers, consolidations, centralization, and headcount reductions:

    a. How has your company, its customers, and its suppliers been affected?

    b. What justifications have been given for headcount reductions (e.g., benchmarking studies, eliminating waste, economies of scale)?

    c. How were the activities administered (e.g., top-down decisions, consultants)?

References

Harrington, H.J., *Total Improvement Management*, McGraw-Hill, New York, 1995, chap. 13.

The Honolulu Advertiser, Bloomberg News Service, Dec. 29, 1998.

Mathews, J. and Katel, P., The cost of quality: faced with hard times, business sours on "Total Quality Management", *Newsweek*, Sept. 7, 1992, pp. 48–49.

2 Employee Involvement in Quality Management

Introduction

The science of quality has grown so dramatically in the last 20 years, that it would be very difficult for anyone in their lifetime to read and mentally process every text written on the subject. In an effort to condense the key quality management elements into two chapters, "Quality Management" is categorized herein as either "employee involvement" or "managerial strategies". While employee involvement is important to the success of every company, this chapter addresses those specific programs in which the employee plays a critical role.

This chapter on employee involvement is organized into two sections: (1) human resources/employee benefits, and (2) success via employee involvement and team building. All of the strategies discussed in this section provide insight on ways that business can improve quality, consistency, and reliability; increase sales; and reduce or eliminate costs. However, when reviewing each with a financially focused mindset, one might realize that providing the best quality products in the world is of little value if the retail price of the product is so high that no one can afford to buy it. In this situation, the product would end up selling at a retail price that is less than the cost of its manufacture. In other words, if the product does not sell, or if the product does not sell with a reasonable profit margin, the company most likely will not be in business long.

When implemented knowingly, "loss leaders" are products offered at a loss in an effort to sell other products at a margin that more than offsets the loss. This approach should be examined with the financial focus to ensure that financially sound decisions are made.

Human Resources

Companies generally employ compensation specialists and analysts whose goal is to establish pay scales that will attract and retain a competent work force. Of course, pay alone should not be the only factor a prospective employee examines before accepting employment with a company. Other factors include:

1. Retirement plans
2. Medical benefits
3. Dental benefits
4. Vacation/holiday benefits
5. Sick pay/salary continuation policy
6. Employee stock options
7. Employee stock ownership plans
8. Tuition reimbursement
9. Flexibility of shifts (working hours)
10. Transportation and commute alternatives
11. Employee recreation facilities (e.g., fitness center, employee store, and activities)
12. Executive perks

Employee Benefits

Companies recognize that everyone is different, and the key to offering a satisfying employment package lies in the appropriate mix of benefits.

Retirement Plan

Retirement plans vary widely from company to company. Whereas some may offer only the traditional Social Security coverage as required by law, others may include company pensions and 401K plans. These will vary, also, and can include such options as company matching to some degree and/or company stock. Many retirement plans are coordinated with large investment institutions, offering employees an assortment of retirement investment options.

Medical and Dental Benefits

Medical benefits again will vary. While some businesses do not have them, those that do often offer a wide assortment of plans, usually at a reduced (company-subsidized) rate. These plans can cover the employee, spouse, and other family members, with the employee's share of the payment being made via payroll deductions.

Vacation/Holiday Benefits

Vacation benefits vary from a simple one week per year to an increasing number of vacation days dependent upon the seniority of the employee. Some companies offer significant sabbaticals (e.g., two months off for special activities) after the employee has worked a predetermined amount of time.

Sick Pay/Salary Continuation Policy

This will vary, depending upon what qualifies as sick time off or time off for personal reasons. Such circumstances may include maternity leave, critical injuries, or serious illness.

Employee Stock Ownership Plans

Many companies offer employees part ownership via employee stock ownership plans (ESOPs), which have the effect of putting the employee in the role of owner. By acquiring shares in the company, the individual reaps the rewards of stock appreciation and dividends. Stock appreciation and dividends are directly related to profits. Because every employee either directly or indirectly contributes to the profit, ESOPs encourage employees to *increase* their contribution to the bottom line. Financial gain is the motivating factor.

An average ESOP allows employees to obtain shares of stock, either by offering a means to buy company stock at a discount or by actually giving shares of stock to employees as part of a savings plan and/or compensation package. There can be significant costs associated with administration of ESOPs, including company matching of employee contributions.

Employee Stock Options

Employee stock options allow employees to increase their income if the company's stock price exceeds a pre-defined level. Here is an example of how an option plan may operate.

On September 1, an employee is given an option to buy 2000 shares of his company for $12 per share. As of September 1, the value of that stock is $12 per share on the open exchange. The employee must hold the stock for at least one year before exercising the option. Exercising the option means he/she can buy 2000 shares for $12 per share. Of course, the employee would not pay $12 per share if he could by it for less on the open market. However, if the value of the company has gone up to $20 per share, the employee would exercise the option. This means he would buy 2000 shares at $12 and then sell the 2000 shares at the current price. The difference between buying and selling price ($20 − $12 = $8) is the profit per share. In this example, the employee receives a bonus of $16,000 by exercising the stock option.

Employees realize that their contribution to the company can affect stock price. An employee with some stock options will often be more motivated than employees without stock options. Because every employee contributes in some way to the value (e.g., stock price) of the company, stock options motivate the employees to strive towards increasing company value. A major positive factor with employee stock options is that the employee does not have to invest a penny unless he or she knows she will make money on the transaction. Again, there can be significant costs associated with administration of stock option plans.

Assistance With Further Education

A number of potential employees just out of high school or college with undergraduate degrees may wish to continue their education; however, for financial reasons, they must seek employment. Such potential employees will look favorably upon a company offering them financial assistance with tuition and flexible work shifts enabling them to pursue higher education.

Company Training Programs

Similarly, an employee may be seeking further education on the job. For example, a computer programmer may seek employment with a company boasting an internal training department that offers employees advanced training in current software applications (e.g., SAP).

Flexibility of Shifts

There are those who, for a myriad of reasons, prefer not to work a traditional 9-to-5 work shift. Alternative shifts may include 4-day work weeks or early morning (e.g., 6:00 a.m. to 3:00 p.m.) or graveyard (e.g., 11:00 p.m. to 7:00 a.m.) shifts. Such flexibility may further enhance the ability of a company to attract employees.

Transportation and Commute Alternatives

Many companies, particularly in large urban areas (for example, New York City or San Francisco), offer employees special benefits to help with their commutes to and from work and between buildings on site. Such benefits include company-sponsored ride sharing (carpooling) coordination and company-owned shuttles that run between local public transportation (e.g., light rail or bus stations) and company offices.

Employee Recreation Facilities

Many larger companies sponsor such recreation features as fitness centers, employee stores, and activities. A company may have an on-site facility which could include an exercise room and shower facilities, or it could work out an agreement so that employees may use the facilities at local health clubs on a discounted basis. Studies have been performed that show company-sponsored health programs almost always pay for themselves in terms of improved employee health (which translates into less sick time). Employee stores have been known to feature discounted laundry, florist, and photoprocessing services and other services such as selling tickets to sports and entertainment events, miscellaneous gifts, and knickknacks (such as clothing featuring the company logo). Some companies are known for rewarding employees by sponsoring elaborate parties (e.g., Christmas parties) and special events (e.g., trips to Hawaii).

Executive Perks

A perk for many in executive management is flying first class on business trips or being allowed to eat in an executive dining room. These are benefits that have very real impacts on the company's bottom line, but companies are willing to pay these perks in order to attract top management talent.

Summary

The above offerings are designed to attract and retain the required work force. These benefits allow employees to satisfy the physiological and safety needs of which Maslow writes. However, once the employees are recruited and working, an effective quality management program works to ensure the employee remains motivated to achieve performance excellence.

Employee Participation in Quality Management Activities

To be successful, many management strategies designed to increase profitability require the active participation of employees. Such programs rely on the experience and performance of knowledgeable employees to achieve significant improvements in operations, which translate to increased profitability.

The success of the following programs, philosophies, and policies relies heavily on employee participation:

1. Total Quality Management
2. Total Quality Service
3. Ownership and Empowerment
4. Employee Suggestion Systems
5. Cost-Reduction or Avoidance Programs
6. Continuous Quality Improvement
7. Process Management
8. Pay for Performance
9. Participative Management and Team Building
10. Quality Circles
11. Task Force Teams

12. Natural Work Teams
13. Process Improvement Teams
14. Effective Meetings
15. Brainstorming
16. Affinity Diagram
17. Lean Process
18. *Kaizen* Event
19. Six Sigma

Psychologists Skinner and Maslow pioneered studies on motivation and behavior manipulation through the satisfaction of various needs. In this section, key motivational techniques and employee-centered programs are presented.

Total Quality Management

The term "Total Quality Management" (TQM) can almost be used synonymously with Continuous Quality Improvement and Process Management. TQM grew from the postwar research and musings of American management consultants such as Armand Fiegenbaum and W. Edwards Deming. It captivated Japanese business leaders in the 1950s and returned to the U.S. in the 1980s. The 1980s saw American business making progress in picking up that quality ball that the Japanese started rolling years before.

Newsweek magazine went so far as to label TQM as the hot boardroom fad of the early 1980s (*Newsweek*, 1992) as millions of corporate dollars were poured into promoting TQM concepts.

A primary focus of TQM is motivating employee involvement and "ownership" of the process they perform. Management should value employees that take an interest in company productivity and take the initiative to seek improvement. Management should realize that before there can be improvement, the potential for improvement must be identified. The individual employee is in the ideal position to note areas for potential improvement.

Total Quality Service

While Total Quality Management has been used as the basis for improvement activities in almost all companies, many enterprises in the service sector have adopted Total Quality Service (TQS) as their program and mission. Similar to TQM, TQS emphasizes many of the current trends in productivity improvement toward the goal of satisfying the customers of its services. Airlines, for example, strive to ensure the traveling public is completely happy with their air travel experience, from the initial process of making reservations to when they pick up their suitcases at the baggage carousel. Employee involvement and ownership are the foundations for TQS.

Empowerment and Ownership

The basic premise of ownership and empowerment is that sharing company ownership and meaningful involvement with employees is a fair and effective means of motivating the work force and achieving many business objectives. In this case, we are not talking about stock ownership. So, how do employees take ownership? How are employees empowered to make significant improvements in work processes? The elements necessary for successful ownership include the following:

1. *Training:* Employees are trained to do their jobs, trained to accomplish their missions, trained to the point where they can be trusted to perform their missions when they are unsupervised.
2. *Environment:* An environment (working area and climate) is created and maintained. Supervisors, managers, and leaders must understand that environment and thrive in it. They can trust the people doing the work and allow them the opportunity to come up with bright ideas and suggest changes and ways to do things better.
3. *Equipment:* Tools and techniques for streamlining and eliminating waste are provided.

Empowerment is closely aligned with the concept of ownership, because when you create a working climate comprised of trained people whose supervisors mentor, coach, and facilitate rather than only control and direct, then you can have ownership. Every time there is a properly administered quality improvement team or process action team, you have empowerment in action. The following sections offer a means to empower employees and offer ownership.

Employee Suggestion Systems

Both Process Management and Continuous Quality Improvement recognize the employee suggestion system as a powerful tool to increase employee participation, quality, and productivity. Suggestions involving cost savings that can be easily quantified should be submitted in a cost reduction or cost avoidance program (see Cost Reduction or Avoidance Programs, below). Other programs should be available for those wishing to suggest changes in procedures or policies for which the benefits are not so obvious but positive outcomes could still result. The key elements of suggestion systems are as follows:

1. A steering committee develops policies and arbitrates in the event of conflict. This committee sometimes includes a member of management.
2. Most suggestions concern human resources, policies and procedures, health and safety, facilities, security, and maintenance.
3. A suggestion committee receives and routes suggestions to evaluators, who should be local experts in the areas relevant to the suggestions.
4. This committee also handles publicity, forms, and feedback.
5. Suggestions and their resolution are published periodically.
6. Feedback is given to the employee making the suggestion and should indicate whether the suggestion was accepted or rejected. Rejection should be accompanied by a thorough explanation and a procedure to be followed if the employee is not satisfied with the explanation. An oral debriefing is often useful for resolving disagreements between the individual and the committee. It is critical that feedback be given in a timely manner and that it reflects the fact that every consideration was given to the suggestion. Employees who feel their ideas merely have been given lip service will often withdraw from these programs, and their participation will be difficult to encourage in the future.
7. Suggestions that result in benefits to the company are often recognized with some sort of reward (for example, a plaque or gift certificate) which is presented to the employee in a formal ceremony. Suggestion systems take many forms, and they should not require significant time and resources to administer.

Cost Reduction or Avoidance Programs

The elements of cost reduction or avoidance programs are the same as those for employee suggestions systems, except that cost reduction submittals can be more easily quantified in dollars. The process by which an employee calculates the savings is similar to that used by estimators when pricing proposals (see Chapter 9). Because the amounts of reported savings are often directly related to a monetary award that the employee receives, pricing of savings must be fairly accurate. However, most employees submitting reports to such programs do not have pricing experience.

Employees are better equipped to calculate dollar savings accurately when provided with such resources as:

1. Training sessions conducted by industrial engineers and trained instructors on how to identify and calculate cost reductions.
2. How-to worksheets available with the submittal form; often such forms are available online.
3. Inclusion of rate tables on the worksheets that provide average costs or rates of broad categories of expenses, such as labor rates, costs of various types of computers, and rates for whatever is common to a particular organization.

Continuous Quality Improvement

It has been said that Continuous Quality Improvement (CQI) is an attitude and a philosophy designed to achieve one objective: customer satisfaction. CQI uses the following analyze-and-improve approach:

1. Analyze capabilities and processes.
2. Improve the capabilities and processes.
3. Analyze capabilities and processes.
4. Improve the capabilities and processes.
5. And so on.

CQI recognizes that customer satisfaction is the result of meeting or exceeding customer expectations for quality, scheduling, and cost. The relationship between cost and quality is value, and value is what the customer wants. CQI targets the work processes — how things are done. The basic CQI concept is

1. Empower *all* people in an enterprise to act. Such empowerment can be successful because the actions taken are based on factual knowledge of the processes in which the employees are involved.
2. Implement actions if they are found to result in one or more of the following goals:
    a. Better products ...
    b. Better services ...
    c. Delivered faster ...
    d. At greater profit.

CQI suggests that getting to these goals depends on systematic analysis of what is being done, for whom, and why. Once these three questions are answered, another systematic application of Process Improvement methods begins.

Management in a true CQI environment relies on the participation and teamwork of the employees to prevent errors before they happen rather than on correcting errors after the fact. Attempts by CQI to remove intimidation as a management style allows people to question long-standing policies or procedures and bring up new ideas in the search for higher productivity. Millions of dollars have been expended throughout the world teaching the concepts of and expounding upon CQI benefits. Instances have been noted, however, where the expenses related to programs similar to CQI have exceeded the benefits received.

Process Management

Process Management (PM) shares many of the same basic ideas of Continuous Quality Improvement and Total Quality Management. PM focuses on the following five subjects:

1. Quality
2. Customers
3. Improvement and innovation
4. Leadership and process (not product and results)
5. Cultivating a learning organization

Pay for Performance

This management strategy, which gained popularity in the mid-1990s, encourages employee performance while eliminating cost-of-living adjustments (COLAs) and similar programs not tied to performance. Many such programs include recognition and incentive bonuses, awards of excellence, employee of the year awards, and merit increases commensurate with levels of performance. Companies using Pay for Performance often abolish COLAs in favor of market-based range adjustments and one-time lump sum bonuses. Despite elimination of COLAs, employee morale is very positive, and studies have shown that this strategy often results in significant increases in employee performance. A key factor is that employees learn not to *expect* that annual pay raise, unless they have raised the level at which they perform their job.

Participative Management and Team Building

We've come a long ways from the days of top-down management, when the all-knowing managers allowed only one-way communications with those reporting to them. Employees are valued more today for what they know and can recommend. And, when such employees are on synergistic teams working together, their value increases significantly. In the 1970s and 1980s, with the advent of such participative management tools as Quality Circles, the value of bottom-up communication and team building was recognized.

Quality Circles

A Quality Circle is a group of volunteer employees from different work areas (may include supplier-customer pairs) that voluntarily meet regularly to identify and propose solutions to work-related problems. This is a team and participative management concept that helped Japan excel in the 1970s and 1980s. Japan made a science out of Quality Circles, providing formal training for management and circle facilitators.

Unfortunately, Quality Circles got a bad reputation in the U.S. because the concept was not used correctly. Companies did not provide the required skill training or support necessary for successful implementation. As a result, U.S. management became discouraged when big dollar savings did not materialize. Also, expectations were so high that management wanted Quality Circles to solve problems that management had never been able to overcome.

Despite the problems experienced in America in the 1980s, the team concept and participative management have flourished in the 1990s.

Use of Quality Circles has grown to encompass a greater array of work groups now generally called "work teams". This trend is expected to continue into the 21st century.

Task Force Teams

A task force team is usually assigned responsibility for resolving a high-level issue or problem. They meet for long periods of time and often break up into small sub-task force teams. Examples of task force teams include the following:

1. Accounting task force, to improve the methods used to account for costs
2. Manufacturing task force, to implement a new manufacturing process (for example, MRP2)
3. Human resources task force, to perform analysis and benchmarking to implement a new performance appraisal process
4. Cash management task force, to close out overdue accounts receivable

Natural Work Teams

The Natural Work Team (also termed "Department Improvement Team") is made up of employees in a particular organization reporting to the same manager. The team is often led by the manager or a supervisor and focuses only on problems for which they possess an expertise. Team size should not exceed 10 members. For organizations larger than 10 employees, membership in the actual team that meets should rotate every six months or so. Rotation gives others a chance to participate. These teams usually start by performing an analysis of the functions performed in their organization to determine the following:

1. Their mission
2. Their customer set
3. Methods for measuring success

The team meets for about one hour each week, and organization problems are identified and prioritized. The team then works to resolve problems having the highest priority. A consensus is reached regarding the solution. The manager has the final veto when considering recommended solutions. A veto would be used if an analysis shows that the recommended solution is not cost effective.

When the tools of Financially Focused Quality are in place, the Natural Work Team functions much more effectively. With team members educated in financial concepts, and with cost in the forefront of their minds, very little time will be afforded to analysis of recommendations that are not cost effective.

Process Improvement Teams

Membership in the Process Improvement Team is assigned by management, consultants, or other individuals intimately involved in the process being examined. The purpose of this team is much more narrowly defined than that of other teams, as here the focus is on a specific process.

In meetings scheduled roughly for one to two hours a week for up to six months, these teams will often identify process issues that can be corrected through the use of a task force team, which will work only as long as it takes to resolve the process issue originally targeted by the Process Improvement Team. The Process Improvement Team will disband once all aspects of the process have been analyzed. Usually, 6 to 12 months later, the Process Improvement Team will reconvene to examine the effects of the prior activity and to take a new look at the process with the latest information available.

Effective Meetings

The teams listed above perform much of their work in meetings. Thus, more and more businesses have been putting increased emphasis on training their employees in the art of conducting "effective meetings". An effective meeting increases profitability for the company in two ways:

1. The time spent by employees in meetings is reduced to what is needed and wanted.
2. The time spent between meetings is much more efficient, due to the successful agreements reached during the meetings themselves.

Most workers are very skeptical about meetings. They seem to feel that meetings are a necessary evil that keeps them from getting their work done. Yet, it is almost impossible to have an inter-agency collaboration without meetings. With technology allowing more convenience by way of telephone conference calls and video teleconferencing, meetings are becoming even more frequent, thus increasing the need to ensure they are efficient and effective. Meetings are held for such purposes as:

1.   Building consensus
2.   Fostering inclusion
3.   Moving towards the vision
4.   Getting things done

Here are some generic hints that may help almost anyone responsible for meetings.

1.   Understand the purpose of the meeting in as simple and clear a manner as possible.
2.   Be sure to invite only those that truly need to be in attendance.
3.   Consider performing a pre-meeting inquiry. This is especially useful for the first meeting of a team or task force, or when the team is taking on new responsibilities or new challenges. Ask members to suggest several agenda topics. Ask them to indicate what they would like to get out of the discussion and to provide an estimate of how much time it will take. Also ask them to suggest potential meeting agreements and to share concerns or potential problems.
4.   Have a clear, concise agenda, and stick to it! The agenda is like a road map in that it is a plan to help the group get to where it wants to go. However, it is rare that a meeting unfolds exactly as prepared. In fact, it would be a miracle if it ever did. Either that, or the person in charge was one bossy son of a gun. If a pre-meeting inquiry has been performed, it is an excellent source for helping the meeting planner shape the agenda around what needs to be covered.
5.   Send out a copy of the agenda prior to the meeting, as well as any materials with which the attendees need to be familiar.
6.   Set the meeting for a time and place agreed upon by all members.
7.   Make the meeting space conducive to getting work done.

8. Pay attention to the "little things" (such as how the meeting room is arranged, the room temperature, whether there is coffee or not) to contribute greatly to the success of a meeting

9. Set the agenda at the beginning of the meeting, along with suggested time limits.

10. Introduce each attendee prior to or at the beginning of the meeting. This is excellent as it enables each member to feel ownership for the meeting, the agenda, and the results. If a premeeting inquiry was performed, share with the team how the inquiry shaped the agenda: "Three people wanted to discuss this, five wanted that, so those topics are on the agenda. The following topics were suggested by only one or two people, but we didn't have time on this agenda. We'll try to cover those topics at a future meeting."

11. Consider the needs of those in attendance, and make an effort to meet those needs that are relevant.

12. Pay attention to the order in which things are discussed and the amount of time spent on each. Often there is a tendency to spend a lot of time discussing items that appear early on the agenda and then to rush through several items towards the end. This can be avoided by making the group aware of this tendency, putting the most important items at the beginning of meeting, and sticking closely to the recommended time allotted to each topic on the agenda.

13. Work to see to it that the format and time allocated for the meeting are a good fit with the purpose of the meeting.

14. Consider changing the agenda as necessary as the meeting progresses.

15. Encourage all members, and give them the opportunity to participate. Do not allow the meeting to be dominated by just a few members.

16. Be sure that all meetings include respect, trust, and, when appropriate, humor.

17. Remember that conflict and problems are okay but should be dealt with in an appropriate, safe manner.

18. Strive for a balance among results, team member relationships, and process.

19. Conduct meetings primarily to accomplish things that can only happen when the team is together.

20. At the end of each meeting, review decisions and action responsibilities.
21. Announce the date of, and list tentative agenda items for, the next meeting.
22. Evaluate meetings after the fact along the lines of a "lessons learned" analysis. Ask members to fill out periodic evaluations of the meetings to aid in your evaluation. This will help in remembering to keep doing what seems to work and to discontinue what doesn't.
23. Distribute meeting minutes. Attendees deserve to receive minutes, and this is particularly important for documenting agreed-upon decisions. It also offers the opportunity to remind team members of the next scheduled meeting. With computers and e-mail, meeting minutes are as easy as adding comments to the earlier published agenda. A push of the button sends the minutes to all attendees.
24. Not all agencies use the same calendar or have the same holidays; it might be a good idea to send out a list of upcoming meeting dates.

Meet at the Right Time and Place

Sometimes, one of the most difficult challenges is just finding a time when every member can attend the meeting. For instance, what can be done if there is absolutely no time that every member can meet? While it may be preferred that everyone be there, the meeting can still be scheduled for a date, time, and place when the majority can attend. For those who have conflicts, here are some options:

1. They can assign a delegate to attend.
2. They can participate via conference call.
3. If practical, discussions requiring their participation can be postponed to a future meeting that they can attend.

Once team members are able to synchronize their meeting times, it is often easiest to meet at the same time and day of the week or month. However, one critical flaw of meetings is to hold them just because they are scheduled on a weekly or monthly basis.

It is extremely important for meeting leaders to ensure that there truly is a need for a meeting before allowing one to take place. A simple preliminary query to team members may reveal that the next scheduled meeting can be canceled. If this is the case, canceling the meeting will most likely come as a very pleasant surprise to the team members. They will certainly look upon the meeting as less of a burden in the future, when they realize that it is held only when a substantive agenda is in place.

Team members should be asked to put the dates on their calendars. Though it sounds obvious, many people still schedule only a week at a time, while periodic meetings can occur weekly and monthly for years into the future. By putting the meeting on the employee's schedule at least two months in advance, the chances of scheduling a conflicting meeting are greatly reduced. As the technology behind such devices as business organizing and scheduling systems, personal digital assistants, Palm Pilots, and personal viewers continues to improve and become less costly, people are now able to schedule ongoing weekly and monthly meetings very easily.

Meeting Locations

Many companies have rooms that are specifically constructed for meetings. For example, an executive conference room may come equipped with the following:

- Mahogany lectern ...
- On a podium ...
- With microphone or lavaliere mike
- Microphones for all attendees
- Special light with dimmers to make the room conducive to viewing presentations
- Facilities for presenting visual information, including: (1) a large tablet, (2) an overhead (Vu-Foil) projector, (3) a slide projector, (4) a computer projection system, and (5) laser pointers

Possible Roles of Attendees

1. *Neutral facilitator*, who takes care of the process of the meeting; a neutral facilitator has no stake in the outcome or content of the meeting. A neutral facilitator can free members to deal with the content of the meeting and levels the playing field.

2. *Facilitating leader*, who is one of the team members trained in facilitation and takes turns facilitating the meetings; they are not completely neutral in regard to the content.

3. *Chairperson*, who is a team member, often elected, who leads the meeting.

4. *Recorder (rarely used)*, who writes down what members are saying on large pieces of easel paper. Care should be taken to write down the actual words and not to paraphrase. Members are responsible for making sure the recorder has captured the essence of what was said.

5. *Secretary/note-taker*, who is responsible for preparing and distributing the minutes of the meeting.

6. *Timekeeper (rarely used)*, who is given responsibility for keeping the meeting flowing in line with the agenda's recommended timetable. If for some reason the person leading the meeting cannot keep track of the time, the timekeeper is given that responsibility.

7. *Participants* (self-explanatory).

8. *Observers*, who are folks interested in the proceedings of the meeting and are allowed to attend, but who had better not say anything (just kidding!).

9. *Set-up people*, who are responsible for arranging chairs, setting up the audio/visual equipment, etc.

10. *Clean-up people*, who need to restore the room to its original condition (e.g., erase the blackboard).

11. *Snack providers* (self-explanatory, though extremely rare in the corporate world).

12. *Aspirin providers* (self-explanatory).

Brainstorming

A valuable component of effective meetings is the use of brainstorming. This is a process that uses more than one person to stimulate the generation of ideas focused toward a particular goal. Brainstorming can be used effectively in a multitude of situations by almost any team (for example, Task Force Teams, Quality Circles, Natural Work Teams, and Process Improvement Teams). Brainstorming can be a great way to use the creativity of the team.

It is imperative that during brainstorming there be no evaluation. Usually, very early in a brainstorming session someone will evaluate an idea. If the "no evaluation" rule is not restated and enforced at that time, the brainstorming will be significantly hampered and participation will quickly cease. Chapter 11 of this text presents the detailed procedure for using brainstorming in Financially Focused Quality.

Affinity Diagrams

The Affinity Diagram takes brainstorming one step further. It is a team process tool that organizes ideas created through brainstorming into natural groupings in a way that stimulates new creative ideas. Without speaking to one another, team members work to generate categories and new ideas. The Affinity Diagram is also known as the KHJ method.

Lean Process

A business process is a group of activities which takes one or more kinds of input and creates an output that has value to an internal or external customer. Examples are writing proposals, software development, and procurement. To "lean" something is to analyze it and remove waste from it, whether it be in wait time, movement from one area to another, or number of approvals required. An enterprise may employ many techniques for "leaning" processes, including *kaizen* events.

Kaizen *Event*

The word *kaizen* (Japanese) means continuous incremental improvement of an activity to create more value with less waste. In a *kaizen* event, a team of affected workers develops a flow of the activity being analyzed and then determine ways to improve it. Barriers are removed, and the team is authorized to implement the necessary changes. There are usually two months of preparation for a *kaizen* event, but only a week is set aside for the actual event.

Six Sigma

Six Sigma is a data-driven method for achieving near-perfect quality. It includes extremely rigorous data gathering and statistical analysis to pinpoint sources of errors and ways to eliminate them. Quality projects are chosen based on customer feedback. This is an effort to improve business by delighting the customer. Utilizing this methodology requires a retraining of the entire workforce, including marketers and janitors, to think and act like engineers. This program has been extremely effective at General Electric, due primarily to the efforts of CEO Jack Welch.

Summary

The management strategies discussed above have been shown to increase employee motivation, which — if channeled properly — offers the prospect of improved profitability. This prospect is enhanced when Financially Focused Quality is used to evaluate the true costs and benefits during each step of the process.

Self-Study/Discussion Questions

1. In what ways do company-sponsored recreation programs foster feelings of love and belonging? Give examples.
2. Each person is different. Which of the factors discussed in the chapter is most important to you when making a job or career change? Which factors would indeed motivate you to work harder and to cooperate more fully with your management?
3. What other options, not mentioned in this chapter, would make you a happier employee? What other options do you think would please your employees?
4. How has your company attempted to implement the following employee involvement techniques:
   a. Empowerment?
   b. Employee suggestion or cost avoidance programs?
   c. Quality Circle, Participative Management, Team Building?
   d. Other?
5. Share some horror stories regarding meetings at your company. Share some successes.

References

Elliott, J.E., *The Philosophy of Process Management,* Lockheed Martin Missiles & Space, Sunnvale, CA, 1993, p. 3.

Mathews, J. and Katel, P., The cost of quality: faced with hard times, business sours on "Total Quality Management", *Newsweek,* Sept. 7, 1992, pp. 48–49.

3 Managerial Strategies for Quality Management

Introduction

The previous chapter discussed quality management programs designed to improve operations via employee involvement. This and the next chapter focus on those quality management programs that are not employee driven. While the employee is almost always involved, it is management, staff, or consultants that take the lead in the programs presented in the next two chapters.

This chapter reviews the following managerial strategies for increasing productivity:

1. Total Improvement Management (TIM)
2. FFQ Blueprint
3. Manufacturing Resource Planning (MRP2)
4. Just-in-Time
5. Business Objectives
6. Vision Statements
7. Mission Statements
8. Benchmarking/Best Practices
9. Training
10. Measurement Processes
11. Core Competencies
12. Outsourcing
13. Reengineering

14. Restructuring
15. Rightsizing
16. Downsizing and Budget Cutting
17. Partnerships With Suppliers
18. Quality Costs and Poor Quality Costs
19. Point of Diminishing Returns
20. Full-Cycle Corrective Action
21. Cost as an Independent Variable (CAIV)

Total Improvement Management

Total Improvement Management (TIM) was originated by a team of authors led by H. James Harrington, International Quality Advisor for Ernst & Young. TIM blends elements of the methodologies of Total Quality Management, Total Productivity Management, Total Cost Management, Total Resource Management, Total Technology Management, and Total Business Management.

At the heart of Harrington's model is the TIM pyramid, arranged as follows:

- *Tier 1: Direction* — top management leadership, business plans, environmental change plans, external customer focus, and quality management systems
- *Tier 2: Basic concepts* — management participation, team building, individual excellence, and supplier relations
- *Tier 3: Delivery process* — process breakthrough, product processes, and service process
- *Tier 4: Organizational impact* — measurements, organizational structure
- *Tier 5: Rewards and recognition*

The Financially Focused Quality Blueprint

Financially Focused Quality adds a new element to the TIM pyramid, creating a modified design called the FFQ Blueprint for total enterprise success (see Figure 3.1). The major addition is educational training and other FFQ tools to ensure that all improvement activities are performed

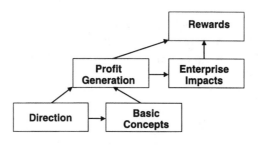

Figure 3.1. The Financially Focused Quality Blueprint

using the Financially Focused Quality mindset. In the middle of the TIM pyramid are the processes by which products are delivered, a component which the FFQ Blueprint labels as "Profit Generation". After all, the processes can run as beautifully as possible, but if they do not generate profits, the business cannot operate for very long. Here is a brief description of FFQ Blueprint components:

Component 1: Direction

The primary purpose of a for-profit enterprise is to make money. That is the number one priority. Top management must establish systems and exercise the leadership skills to accomplish this purpose. Direction is exhibited in the following five elements.

1. *Management leadership:* Leading is more than supplying the resources for the latest improvement process. It includes participating in the design and being part of the process.
2. *Business plans:* Giving direction requires that business plans be completely thought out and communicated clearly to all employees.
3. *Environmental change plans:* Valid implementations of new processes will usually require changing the work environment. Successful FFQ necessitates a financially focused environment, achieved via financial training.
4. *External customer focus:* Profitable enterprises have an excellent understanding of, and a close working relationship with, their external customer/consumer.

5. *Quality management systems* (for example, ISO 9000): Harrington calls these systems the "blocking and tackling" of the improvement process and the essential building blocks for the rest of the structure.

Component 2: Basic Concepts

1. *Management participation:* Management's active participation will be facilitated when management feels comfortable in its leadership role.
2. *Team building:* Synergies are achieved when employees unite in teams with common goals.
3. *Individual excellence:* Each employee must be motivated, empowered, and equipped with the proper tools (for example, financial training) to make a difference and to spark positive impacts.
4. *Supplier relations:* Both the organization and its supply organizations should work together to achieve mutual success.

Component 3: Profit Generation

Notice that this is the only component that is linked to all the others. Profit generation is the central theme. The elements contained in this component are designed to generate profit through delivery of products and services.

1. *Process breakthrough:* Benchmarking, Process Improvement Teams, Just-in-Time Manufacturing, and Outsourcing are just a few of the approaches utilized to improve processes and the output that customers receive.
2. *Product processes:* The process in place to deliver products must operate at a peak level of excellence. This means the processes themselves are well designed and documented, and effort is continually undertaken to improve the product delivery process using a financially focused mindset.
3. *Service process:* The delivery process for services varies from the process for delivery of products. This element focuses on the design, documentation, and continual improvement of the service delivery process using a financially focused mindset.

Component 4: Enterprise Impacts

With the improvement process well underway, the next component of the FFQ Blueprint is addressing how changes impact the enterprise. There are two elements to examine:

1. *Measurements:* Only when the improvement process documents positive measurable results can management be expected to embrace the newly implemented methodology as a way of life. This measurement process should apply the basic financial concepts touted by FFQ.
2. *Organizational structure:* Large organizations need to give way to small business units that can react quickly and effectively to changing business needs.

Component 5: Rewards and Recognition

This component is the result of a well laid out FFQ Blueprint. The employees are rewarded with job security and good pay. Employees and teams may also receive recognition with special awards and pats on the back. Company owners are rewarded with increasing dividends, profits, and stock value.

Manufacturing Resource Planning

In an effort to improve efficiencies, many manufacturing companies with adequate computer systems have been implementing the Manufacturing Resource Planning (MRP2) methodology. MRP2 is a closed-loop system integrating and managing all the resources used in the production of products and/or services. For complex manufacturing concerns, a fully functional MRP2 capability requires computer software containing most if not all of the following modules:

1. Bill of Materials (BOM): Those items called out by the engineering drawings to build the end product listing all the sub-assembly parts and raw materials that go into a parent assembly, showing the quantity of each required to make an assembly. This can also include the packing BOM.

2. Machine capacity planning (MCP; the maximum output a machine could process)
3. Manpower planning
4. Production scheduler
5. Work-in-progress tracking (WIP)
6. Production quality control
7. Production tracking
8. Finished goods (FG) inventory
9. System utilities

Just-in-Time Manufacturing

Just-in-Time (JIT) is a manufacturing philosophy that is designed to produce the required items, at the required quality, in the right quantities, at the exact time they are needed. The goal of this approach is to achieve excellence and eliminate waste. The waste referred to in this context is all things that do not add value to the product (e.g., overproduction, inventory, defective products, transport and waiting time).

Just-in-Time integration can be found in four areas of the manufacturing firm: accounting, engineering, customer, and supplier. On the accounting side, JIT relates to work-in-process (WIP), utilization, and overhead allocation. The engineering side of JIT focuses on simultaneous and participative design of products and processes.

Business Objectives

The direction for the organization over a period of time is set by business objectives. For example, organizations may establish objectives to complete in the following year. Performance in relation to these goals is regularly monitored. It is not uncommon for management to set goals that are easily achieved. For such objectives to be meaningful, however, they should be challenging.

Vision Statements

Many companies have been adopting Vision Statements to help focus all employees in the same direction. A Vision Statement presents a view of what the company should be like 10 or 20 years into the future. It could be as simple as that established by the Des Moines City Government: "*A friendly and safe waterfront community.*"

A Vision Statement can also be more detailed, such as this one for the California State University, Monterey Bay:

> "*California State University, Monterey Bay (CSUMB) is envisioned as a comprehensive state university, which values service through high-quality education. The campus will be distinctive in serving the diverse people of California, especially the working class and historically under-educated and low-income populations. It will feature an enriched living and learning environment and year-round operation. The identity of the University will be framed by substantive commitment to a multilingual, multicultural, intellectual community distinguished by partnerships with existing institutions, both public and private, and by cooperative agreements which enable students, faculty, and staff to cross institutional boundaries for innovative instruction, broadly defined scholarly and creative activity, and coordinated community service.*
>
> "*The University will invest in preparation for the future through integrated and experimental use of technologies as resources to people, catalysts for learning, and providers of increased access and enriched quality learning. The curricula of CSUMB will be student and society-centered and of sufficient breadth and depth to meet statewide and regional needs, specifically those involving both inner-city and isolated rural populations (Monterey, Santa Cruz, and San Benito). The programs of instruction will strive for distinction, building on regional assets in developing specialty clusters in such areas as the sciences (marine, atmospheric, and environmental); visual and performing arts and related humanities; language, culture, and international studies; education; business; studies of human behavior, information, and communication, within broad curricular areas; and professional study.*
>
> "*The University will develop a culture of innovation in its overall conceptual design and organization, and will utilize new and varied pedagogical and instructional approaches including distance learning. Institutional programs will value and cultivate creative and productive talents of students, faculty, and staff, and seek ways to contribute to the economy of the state, the well-being of our communities, and the quality of life and development of its students, faculty, and service areas.*
>
> "*The education programs at CSUMB will*
>
> ■ *Integrate the sciences and the arts and humanities, liberal studies, and professional training;*

- *Integrate modern learning technology and pedagogy to create liberal education adequate for the contemporary world;*
- *Integrate work and learning, service and reflection;*
- *Recognize the importance of global interdependence;*
- *Invest in languages and cross-cultural competence; and*
- *Emphasize those topics most central to the local area's economy and accessible residential learning environment.*

"The University will provide a new model of organizing, managing, and financing higher education:

"The University will be integrated with other institutions, essentially collaborative in its orientation, and active in seeking partnerships across institutional boundaries. It will develop and implement various arrangements for sharing courses, curriculum, faculty, students, and facilities with other institutions. The organizational structure of the University will reflect a belief in the importance of each administrative staff and faculty member, working to integrate the university community across 'staff' and 'faculty' lines.

"The financial aid system will emphasize a fundamental commitment to equity and access. The budget and financial systems, including student fees, will provide for efficient and effective operation of the University.

"University governance will be exercised with a substantial amount of autonomy and independence within a very broad CSU system-wide policy context.

"Accountability will emphasize careful evaluation and assessment of results and student learning goals.

"Our vision of the goals of California State University, Monterey Bay includes a model, pluralistic, academic community where all learn and teach one another in an atmosphere of mutual respect and pursuit of excellence; a faculty and staff motivated to excel in their respective fields as well as to contribute to the broadly defined university environment. Our graduates will have an understanding of interdependence and global competence, distinctive technical and educational skills, the experience and abilities to contribute to California's high-quality work force, the critical thinking abilities to be productive citizens, and the social responsibility and skills to be community builders. CSUMB will dynamically link the past, present, and future by responding to historical and changing conditions, experimenting with strategies which increase access, improving quality, and lowering costs through education in a distinctive CSU environment.

"University students and personnel will attempt analytically and creatively to meet critical state and regional needs and to provide California with responsible and creative leadership for the global 21st century."

The California State University Vision Statement can actually be broken down into several Vision Statements, with different statements applying to different departments and organizations within the university.

Mission Statement

The Mission Statement is very similar to the Vision Statement, and much overlap exists. The Mission Statement can also be described as a purpose statement and frequently will not be as detailed as the California State University Vision Statement above. For example (and for contrast), the Mission Statement of the Des Moines City Government is *"To enrich residential living by providing leadership, administration, and community services that reflect the pride and values of Des Moines."*

The Mission Statement for hamburger giant McDonalds is *"To satisfy the world's appetite for good food, well served, at a price people can afford."*

Benchmarking/Best Practices

Benchmarking is the process of identifying, learning, and adapting outstanding practices and processes from any organization, anywhere in the world, to help an organization improve its performance. The primary purpose of the benchmarking process is to gather the tacit knowledge — the know-how, judgment, and enablers — that explicit knowledge often misses. It should be noted, however, that benchmarking is not the same as benchmarks. Benchmarks are performance measures: How fast? How many? How close? How far? Benchmarking, on the other hand, is action — discovering the specific processes and practices in place in a successful operation that are responsible for high performance, understanding how these practices work, and adapting and applying them to your own organization. While benchmarks are facts that enable tracking performance to pre-determined goals, benchmarking enables real improvement.

Benchmarking leads to identification of best practices. No single "best practice" exists because what is best for one company is not necessarily best for all. Every organization differs from others in some way. That is because organizations usually have a unique blend of cultures, environments, missions, and technologies. When discussing "best", we are referring to those practices that have been proven to produce superior outcomes, selected by a consistent and systematic process, and evaluated and rated at the top for all practices examined.

Below is a listing of some of the positive factors an organization may experience as a result of benchmarking. Benchmarking can:

1. Prevent reinventing the wheel. A company should not invest the time and expense when another enterprise may have done it already — and often better, more cheaply, and faster.
2. Accelerate change and restructuring by taking advantage of tested and proven practices.
3. Convince skeptics who can see that a new process works.
4. Help overcome inertia and complacency and help create a sense of urgency when gaps are revealed.
5. Lead to "outside the box" ideas by looking to other industries for ways to improve.
6. Require that organizations examine present processes which often leads to improvement in and of itself.
7. Make implementation more likely because the process owners are usually involved in the benchmarking process itself.

Training

In the previous chapter, we discussed motivation theory and the role motivation plays in relation to productivity. There is no question about the value of having a motivated work force. Motivated employees will work harder, faster, and with more care and pride. But, motivation does not always lead to working *smarter*. Towards the goal of working smarter, many motivated employees will pursue further education on their own. They may join professional societies, read technical journals, or return to school for formal education. However, for a myriad of reasons, not all employees (even those that are highly motivated) are willing to undertake additional studies on their own time and at their own expense. The only way a company can ensure that its workers are educated in regard to the latest special process to increase productivity is to sponsor such courses on company time and with company resources.

Company-sponsored training comes in many forms, including:

1. Formal classroom training
2. Self-pace study via workbook, computer, or other devices

3. Personalized trainers and consultants (this category can include telephone assistance — for example, a centralized help desk — or trainers who meet with employees as needed)

Content of such training may include:

1. How to use the latest software licensed at the facility (e.g., *Microsoft Office 98*)
2. Current strategies for improved profitability (e.g., Full-Cycle Corrective Action)
3. Programming languages (e.g., Java)
4. Management training (e.g., "Introduction to Supervision")
5. Topics unique to a specific enterprise (e.g., "How To Close a Sale")
6. Effective presentations
7. Effective meetings
8. Employee orientation
9. First aid/CPR
10. Low-voltage electrical safety
11. Hearing conservation

Depending upon the size of the company, the list of topics could go on and on.

One unfortunate aspect of the nature of corporate training departments is that when overhead budgets get slashed training budgets are some of the first to get hit. It is important to utilize a financial focus when determining which classes to cut. The tools of Financially Focused Quality can help greatly in making such a determination.

Measurement Process

Any quality management process that requires a significant investment in time and or money should also have a comprehensive measurement plan. The organization should develop a plan that demonstrates how the affected goals are met. These goals should include quality and productivity, but the main focus of these plans should be profit.

If the measurement process does, indeed, document favorable results, management can be expected to adopt the new methodology as an improved way to do business. A good measurement plan can get someone who has been "sitting on the fence" to jump onto the side of the related quality management process. As the process continues and develops further, it becomes a part of the overall process. Thus, the measurement methodology can be modified, and eventually a separate measurement plan will no longer be required.

Core Competencies

Looking through any number of annual reports of *Fortune* 500 companies, you will find the familiar jargon "focusing on core competencies". This term is used by companies that at one time diversified by developing new lines of business. Many of these companies came to a realization that they should not be looking for new businesses, but rather should be working to improve upon their original money makers, also known as "core competencies".

Outsourcing

When an enterprise is just getting started, a small staff can often handle most of the activities. Take, for example, a small company that manufactures handmade notepaper. A staff of two or three people can most likely handle all business operations on their own. They will market, design, make, package, and mail their products and catalogs.

As the company grows, however, these people will need to hire more people for specific activities. It is conceivable that some day they may have a staff of five employees who just take telephone orders for 8 hours each day. Or a staff of 15 employees who spend each day making notepaper by hand. Or two employees in the mailroom packaging and mailing the product.

Many companies find that it is cheaper to outsource, which means to contract to have some of their work performed by another company. In the above example, Company A may be retained to handle telephone orders. Company A represents 100 manufacturers and has over 150 people taking telephone orders. The resulting economies of scale allow the notepaper company to save a considerable amount of money. Five employees are removed from payroll, and costs associated with the phone order system are eliminated. However, when outsourcing takes place and employees leave the payroll, there are fewer employers to which overhead costs may be allocated. That means that the overhead costs associated with each of the remaining employees will most likely increase. Companies need to consider all financial ramifications when making outsourcing decisions. See Case Study D for further analysis.

Reengineering

Reengineering restates many of the quality management philosophies. Its primary drive is to understand and improve processes. Toward this goal, it examines the following:

1. Critical process characteristics
2. Effectiveness
3. Efficiency
4. Processing time vs. cycle time
5. The cost cycle-time chart
6. Streamlining the process
7. Measurements, feedback, and action
8. Data collection
9. Variables vs. attributes
10. Describing and eliminating variation distributions
11. Quality improvement tools
12. Control charts used in business process improvement

Systems Reengineering is used to sustain and enhance existing computer and information systems for migration to new and more advanced environments. This methodology reengineers the functional logic of an existing system into reusable modules. By reusing the best parts of a system, new systems development is avoided and systems investments are protected.

Restructuring

Restructuring is a term that is sometimes used concurrently with reorganizing, and it almost always means headcount reductions. The term is used to define the process in which an enterprise takes steps to become more efficient through consolidation, merger, or dissolution of organizations. In one such strategy, a company may originally be organized with eight main divisions, as follows:

1. Manufacturing
2. Quality Assurance
3. Engineering
4. Marketing
5. Finance
6. Human Resources
7. Logistic
8. Procurement

Restructuring by business units would have the company grouped into just four divisions, organized by the products they provide both externally and internally, as follows:

1. Compact Disc Division (external customers)
2. Audio Tape Division (external customers)
3. General Operations, including marketing, finance, human resources (internal customers)
4. Engineering (internal customers)

Such restructuring can result in big personnel reductions. For starters, if each of the eight divisions above had a director, reducing to four business units could reduce the number of directors and their staffs by 50%.

Rightsizing/Smartsizing

Rightsizing (a.k.a. smartsizing) is similar to restructuring and reengineering. The goal here is to employ the "right" number of employees to get the job done as efficiently as possible. This usually means eliminating or combining similar tasks, leading to load leveling. When work loads are leveled, all employees are working at similar output levels. Benchmarking studies are frequently performed as a basis for rightsizing.

Downsizing and Budget Cutting

Other strategies still used by industry to achieve change are budget cutting and downsizing. Restructuring usually leads to headcount reductions. Reengineering generally leads to headcount reductions. Rightsizing usually leads to headcount reductions. Then there's downsizing (a.k.a. capsizing?), which has several drawbacks:

1. Change is so fundamental that it should be performed to correspond to what is happening in the relevant industry. Simply removing resources alone without changing the underlying process usually is self-defeating.
2. Downsizing is counter to a company's culture of long-term employment. Any management action that results in headcount reductions can have a significant impact on employee morale.

Partnerships With Suppliers

Many companies have learned that it is not always beneficial to shop around for the lowest price when seeking suppliers. Often, suppliers will bid low to get new business. With new suppliers, a learning curves take place, and often total costs with a new supplier can exceed those associated with the original supplier, when one considers the additional effort necessary to break in a new supplier.

Establishing a special partnering relationship with a supplier benefits both parties:

1. Suppliers will work extra hard and pull strings (for example, expedite schedules) for a company that has been its bread and butter in past years.
2. Companies can alert suppliers to new trends and requirements in advance, allowing for smooth transitions to new processes.
3. The longer a supplier and company work together, the more the learning curve allows processes to become fine-tuned. Procurement system refinements result in a better bottom line for both companies.

Quality Costs and Poor Quality Costs

Quality Cost is one of the approaches used for quantifying the value of quality assurance measures in a manufacturing environment. The four basic Quality Costs are

1. Prevention cost (Quality Cost)
2. Appraisal cost (Quality Cost)
3. Internal cost of failures (Poor Quality Cost)
4. External cost of failures (Poor Quality Cost)

Full-Cycle Corrective Action

Full-Cycle Corrective Action (FCCA) was introduced to the quality world in the early 1980s. FCCA took Quality Cost one step further by introducing tools to examine the total Quality Cost trade-off in correcting manufacturing defects. Much of the framework for Financially Focused Quality was established in the pioneering works on FCCA.

Cost as an Independent Variable

The Department of Defense's Cost as an Independent Variable (CAIV) is an important principle of acquisition reform. While FCCA tools were first applied to manufacturing, CAIV targets costs over the entire life-cycle of the product. Cost trade-offs are examined by a Cost Performance Integrated Product Team (CPIPT) during all phases of the acquisition program. The CAIV process also involves execution of the program in such a way as to meet or reduce stated cost objectives. Organization and activities of the CPIPT are described in DoD 5000.2-R.

Summary

All of these approaches and processes are designed to improve operations, which should lead to increased profitability. However, what tools are built into these mechanisms to ensure a profit? Only Quality Costs Full-Cycle Corrective Action, and Cost as an Independent Variable offer a financial perspective. Financially Focused Quality offers the system to put a financial focus on every potential quality enhancement and cost-saving measure. The only approach to quality that will succeed is the one that elevates finance to equal footing with quality.

Self-Study/Discussion Questions

1. Which of the quality management approaches do you think could be the most expensive to implement? Could be the most expensive to maintain? Could produce the most benefit measured in savings? Could produce the most benefit measured in employee morale?

2. What techniques almost always lead to headcount reductions? What are some of the drawbacks of this strategy? How can a company measure the impact of lower employee moral?

3. Are there some examples in your life (work or personal) where excess preventative costs are paid? How do you feel about paying for extended service agreements on products such as washers, dryers, and computers? Why do you think retailers sell such extensions?

References

Cappels, T.M., *Full-Cycle Corrective Action: Managing for Quality and Profits,* Quality Press, Milwaukee, WI, 1994.

Elliott, J.E., *The Philosophy of Process Management,* Lockheed Martin Missiles & Space, Sunnyvale, CA, 1992, p. 3.

Harrington, H.J., *Total Improvement Management,* McGraw-Hill, New York, 1995, pp. 32–38.

Mathews, J. and Katel, P., The cost of quality: faced with hard times, business sours on "Total Quality Management", *Newsweek,* Sept. 7, 1992, pp. 48–49.

4 Quality Management Systems and Customer Focus

Introduction

Quality management systems are designed to ensure that an organization consistently meets customer requirements. They define how a company should operate to achieve this goal. Management teams recognize that without customers there would be no business. Satisfying the customer while maintaining a customer focus is a science unto itself. Also, standardizing business processes and products enhances business operations by increasing efficiency and improving customer satisfaction. The international standardization of processes and products is another major challenge.

This chapter reviews the following managerial strategies for increasing productivity:

1. Quality management systems
2. Standardization
3. ISO 9000
4. ISO 14000
5. External customer focus
6. Customer feedback

Quality Management Systems

Though quality management systems take many forms, they hold the following elements in common:

- They cover a wide variety of the organization's activities. Definitions of quality are expressed in nonconfining terminology that includes not only product performance characteristics but also the service characteristics (for example, after-market support) that the customer demands.
- They focus on creating and maintaining a consistent work process. This often requires documentation of the work process in an effort to achieve standardization.
- They emphasize the prevention of errors, rather than relying on error detection and reaction after the fact. At the same time, however, quality management systems also emphasize a corrective action process for those errors that do occur. This process attempts to be closed loop, in that it includes detection, feedback, and correction of the error. When the system is administered effectively, the error should not recur.
- They include measurement mechanisms to demonstrate their effectiveness and/or identify problem areas.

Standardization

In today's global economies, steps toward standardization are frequently being taken. Standard currencies for groups of nations are being developed. Standard codes are used by various airlines to designate airports. Languages and units of measurement are also targets for standardization.

The Euro

On January 1, 1999, the dream of a European monetary union became a reality when 11 European nations combined their currencies into one, fixing and announcing their national exchange rates against a new currency called the *euro.*

The French franc, German mark, Italian lira, and eight other currencies became unalterable fractions of the euro as Europe forged a vital new chapter in its drive toward economic integration.

Movie, Video, and Computer Products

The following events in the history of movie, video, and computer product advancements show how emerging technologies are impacted by the lack of global standards.

8-mm Movies: Super 8 vs. Single 8 Movie Formats

In the 1950s and 1960s, many families were using 8-mm (millimeter) home movie equipment to record lasting memories on 8-mm film. In the late 1960s, two companies developed a way to fit a bigger image on film that was 8 mm wide. This resulted in brighter, clearer images when projected onto a screen. Eastman Kodak developed "Super 8" film, which came in a cartridge that could easily be loaded into a Super 8 camera. Fuji developed "Single 8" film, which also came in a cartridge. However, the Single 8 format offered other benefits to movie buffs because it allowed the film to be rewound inside the cartridge. This feature allowed filmmakers to create double exposures, a technique necessary for such special effects as super-imposed titles.

Eastman Kodak's dominance in the U.S. eventually led to the Super 8 format being victorious. In fact, many consumers in the U.S. were never even aware of the Single 8 system or its advantages. Had Kodak and Fuji worked together to develop a standardized format, Single 8 may have been the victor, and everyone would have benefited

Beta vs. VHS Videotape Formats

The 1970s and 1980s saw the world changing again, as videotape recorders became readily available to the public. Two formats were again at battle. Sony Corporation developed what many claimed was a superior format which was known as Beta. A group of other manufacturers released their version of videotape known as VHS. Despite the advantages of Beta, VHS won this battle and is still prevalent as we move into the 21st century.

VHS vs. 8-mm Videotape

Sony scaled back on selling its Beta format and enjoyed much success after developing and marketing its 8-mm videotape format. 8-mm videotape offered many advantages over VHS — in particular, better sound quality and a smaller, more convenient size.

The VHS manufacturers countered with VHS-C — VHS tape that was in a much smaller cartridge and comparable in size to Sony's 8-mm videotape. Sony however, changed the game plan that they had used in the days of Beta. Sony allowed other companies to manufacture 8-mm tape. As a result of this strategy, 8-mm video systems are still around today and do not show any sign of going the way of Beta.

Macintosh vs. Personal Computers

In a manner similar to the Beta vs. VHS battle, personal computers have been the subject of a tremendous battle waged between the Macintosh and the PC formats. After the Commodore and Radio Shack computers appeared, real competition began to take place between IBM and a new start-up company known as Apple. The IBM personal computer (PC) could do a lot for its owners, but using one required typing in sometimes complicated codes and commands to get the computer to do anything.

Then Apple developed their unique point-and-click desktop computer. It was easy to use and very popular, but Apple would not license any other companies to manufacture the product. Another new company, Microsoft, developed their "Windows" technology. This software, when loaded onto a PC, offered a similar point-and-click capability that, up to that time, had only been available on the Apple products. By this time, many other companies were manufacturing IBM-type computers, and the Microsoft software made them all very popular.

With the great number of PCs being sold, the market for PC support was expanding much faster than that for Macs. As a result, software manufacturers and computer peripheral manufacturers began focusing on products for PCs. This led to a greater assortment and lower prices for PC products than for Macintosh products. Many in the industry will say that the Mac is a superior computer that offers a better operating system than the PC; however, the PC standardization of most manufacturers had a very serious impact, as Apple lost a great deal of market share in the 1990s.

ISO 9000 and ISO 14000 Series

In the world of international commerce, business survival depends upon many factors, including free trade across borders, which allows competition on equal terms, thus reducing costs and improving profitability, efficiency, and the company's image. The worldwide free movement of goods and services without trade barriers is, of course, the wish of every manufacturer and supplier.

Standards have, for a long time, been closely associated with trade. Agreements on formal standardization are making life easier for buyers and sellers of goods and services around the world.

ISO 9000 Series

Markets are now becoming global, and supply chains can cross many borders. International agreements and standards, such as ISO 9000 and ISO 14000, can facilitate this cross-border trade. These standards are established by the International Organization for Standardization (ISO). The U.S. member of ISO is the American National Standards Institute (ANSI).

The ISO 9000 series is a set of five individual, but related, documents that define international standards for quality management systems. These documents define the elements an organization would need in order to implement and maintain an effective quality management system:

1. *ISO 9000:* This document serves as the legend for the entire series. It provides the user with guidelines for selecting and using ISO 9001, 9002, 9003, and 9004.
2. *ISO 9001, 9002, and 9003:* These are quality system models for external quality assurance. These three models, which are subsets of each other, define the specific tasks the company performs to generate income. It is against these "contracted" requirements that the company is audited. *ISO 9001* is the most comprehensive standard in the series and covers design, manufacturing, installation, and servicing systems. *ISO 9002* covers production and installation but does not cover design or service functions. *ISO 9003* covers only the final inspection and test and applies to an organization that resells products it purchases; it is limited in its guidance, because it focuses on inspection and not the manufacturing process itself.
3. *ISO 9004:* This is a guideline for implementing and auditing the actual quality management system.

ISO 14000 Series

The ISO 14000 series is a growing set of documents that focus on international environmental concerns. Environmental credibility has become a factor in national and international competitiveness. Implementation of ISO 14000 standards and subsequent certification can facilitate progress towards increased competitiveness through measurement and innovation, leading to increased profit, more efficient processes, reduced costs, and a more credible image.

The first of the series, ISO 14001, offers a common, harmonized approach for use among all organizations, wherever they are in the world. Designing processes and equipment to include environmental considerations requires an evaluation of all aspects of a product or service (ideally, from "womb to tomb", although this is not explicitly stated by ISO 14001). It is only through the establishment of an Environmental Management System (EMS) that an organization can, over time, monitor and control these aspects. In other words, an effective program of design for the environment requires an EMS.

Establishing and operating an Environmental Management System can be the foundation of an internal risk control program. It helps to ensure that environmental issues are considered strategically, rather than as a special exercise. Environmental management is no longer something extra which organizations need to do for moral or corporate responsibility reasons. Instead, it is a part of every company's business strategy to help achieve a competitive edge. Stricter enforcement of environmental legislation, coupled with heightened awareness of customers and stakeholders in regard to the potential risks associated with environmental liabilities, has led to increased interest in the legal, financial, and commercial risks associated with environmental performance.

A listing of key ISO 14000 series documents follows:

1. *ISO 14001: Environmental Management Systems* — Specifications with guidance for use
2. *ISO 14002: Environmental Management Systems* — Guidelines on ISO 14001 for small and medium-sized enterprises
3. *ISO 14004: Environmental Management Systems* — General guidelines on principles, systems, and supporting techniques
4. *ISO 14010: Guidelines for Environmental Auditing* — General principles

5. *ISO 14011: Guidelines for Environmental Auditing*— Audit procedures/auditing of environmental management systems
6. *ISO 14012: Guidelines for Environmental Auditing*— Qualification criteria for environmental auditors
7. *ISO 14015: Environmental Assessments of Sites and Entities*
8. *ISO 14020: Environmental Labels and Declarations* — General principles
9. *ISO 14021: Environmental Labels and Declarations* — Self-declaration environmental claims/guidelines and definition and usage of terms
10. *ISO 14024: Environmental Labels and Declarations* — Environmental Labeling Type I/guiding principles and procedures
11. *ISO 14025: Environmental Labels and Declarations* — Environmental Labeling Type III/guiding principles and procedures
12. *ISO 14031: Environmental Management* —Environmental performance evaluation/guidelines
13. *ISO 14040: Environmental Management* — Life-cycle assessment/principles and framework
14. *ISO 14041: Environmental Management* — Life-cycle assessment/goal and scope definition and inventory analysis
15. *ISO 14042: Environmental Management* — Life-cycle assessment/life-cycle impact assessment
16. *ISO 14043: Environmental Management* — Life-cycle assessment/life-cycle interpretation
17. *ISO 14050: Environmental Management* — Vocabulary

As the financial and legal risks associated with poor environmental performance have increased and become more quantifiable, other areas of business, notably those dealing with finance, risk management, and insurance, are becoming increasingly involved in the management of environmental performance. Insurance companies, for example, are starting to demand much more detailed information about pollution exposure.

External Customer Focus

In the 1980s, the customer was proclaimed the omnipotent ruler. "The customer is always right" became a catch phrase. Competition to serve the customer became so intense that new performance standards were being developed every day. Harrington (1995) reports that one afternoon Premier Industrial Corporation got a call from Caterpillar informing them that their production line was down because a $10 relay had malfunctioned. Premier's replacement parts were located in a warehouse in Los Angeles. By 10:30 p.m. that evening, Caterpillar's line was up and running again. Service like that costs Premier a lot of money, but it pays off. Premier receives as much as 50% more for its parts, and its return on equity is 28%.

The quality management philosophy has always been very much concerned with satisfying the external customer. A company will know it is on the right track if it can answer "yes" to the following questions:

1. Are your products or services designed to meet customer needs?
2. Do you know whether your business operations are helping or hurting your relationships with your customers?
3. Can you tell whether or not you are satisfying your customers without doing a survey?

Today's companies are striving more than ever to:

1. Design customer-focused products and services
2. Develop customer-focused measurement systems that allow a company to manage proactively how well it is satisfying customer needs
3. Design, redesign, and improve business operations and processes so that they are capable of meeting customer needs

Customer Feedback

Organizations have many ways they can measure how well they are satisfying their customers. The most direct method is to get information directly from them. Customer feedback is any way in which customers' reactions to products or services are relayed back to the product or service suppliers.

Expensive Products and Customer Feedback

When a consumer spends $40,000 for a new car, would he or she be concerned about a tiny dent in the fender or a little scratch in the paint? You better believe it! Any anxiety felt by the consumer would be dramatically compounded when, upon casually reviewing the car before taking final acceptance, a modest scratch is observed. The customer most likely would immediately seek out a representative of the sales department and report this flaw.

Imagine now that a defect has not been noticed prior to acceptance of the car. The consumer has taken receipt of the vehicle and is driving happily along when he notices a light bulb in the glove compartment is burned-out. The odds are that he would drive back to the dealer and have the bulb replaced.

In another example, the consumer, while driving, tries to adjust the rear view mirror and has difficulty getting the mirror to remain in the position that allows maximum visibility. Eventually, the driver realizes that by squeezing the base of the mirror tightly, the mirror finally stays in the desired position. Though it is only a minor annoyance, the consumer might decide to return to the dealer and have the mirror fixture repaired or replace; however, the consumer may rationalize that the effort involved in making such a failure report would be greater than the benefit. As a result of this rationalization, the following occurs:

1. The consumer develops a slight feeling of discontent with the product, vendor, or manufacturer thereof.
2. The manufacturer does not learn of this instance of failure. Lacking notification of such a failure, the chances of correcting it are reduced.

Inexpensive Products and Customer Feedback

Manufacturers of low-cost items have a much lower incidence of failures being reported by customers. For example, an individual may purchase a headphone radio from a mail-order company for a very low price (e.g., $4 plus $3.50 for shipping and handling). Upon receipt of the product, the purchaser may wait a week or two before going to a store to buy the requisite batteries. When the product is tried, the customer finds that the reception is terrible. Only two or three stations can be received, and even then the static interference is very annoying. In this example, the buyer has several options, including the following:

1. Repackage the product and mail it back to the mail-order company for exchange, refund, or credit (depending upon the

company policy). This option would likely cost more in time and postage than the product is worth.

2. Throw away the headphone radio.
3. Try to fix it.
4. Take apart the headphone radio and use the parts for experiments.
5. Give it to a relative or friend and react with surprise when hearing that it doesn't work, as if not aware of the inherent problems of the headphone radio.

Of these options, the only one that provides feedback is returning the product to the vendor, assuming it is still in business. Even then, there is the possibility that the company that manufactured the radio will not learn of this specific failure.

Service Industry Feedback — Surveys and Questionnaires

The service industry is heavily dependent upon customer feedback for determining their successes and failures; therefore, it is not uncommon for customer questionnaires to be found in hotel rooms and in restaurants.

Service Industry Feedback — Real Time (Restaurant)

A second feedback mechanism is the first person, immediate report of a problem. Entertainers Tom and Bob offer the following example of feedback in a restaurant (from Roll and Cappels, 1973):

The customer, an obviously well-to-do gentleman, is seated at a nice table in an elegant restaurant in a posh San Francisco hotel. He is sipping a Manhattan (on the rocks) as he carefully reviews the menu. A quartet of musicians is playing some light contemporary jazz as the customer sets his menu down. Within seconds, a tuxedoed waiter is at the table, prepared to take the order.

The customer — a middle-aged man in a Giorgio Armani suit — orders the prime rib "early bird" special. The waiter nods approvingly at the selection and inquires as to how the meat should be cooked.

"Medium rare," replies the gentleman.

"And, sir," continues the waiter, "You have a choice of soup or salad. The soup today is split pea."

"Well, how about that?" says the customer. "I feel like split pea soup today!"

"That's funny," remarks the waiter. "You don't look like split pea soup!"

The customer is mildly amused. "Very droll. Now please don't dally."

"I'm outta here," the waiter responds. "I'll be right back with your soup." The waiter waltzes off towards the kitchen.

Taking another sip of his Manhattan, the gentleman loosens his tie and settles back into his chair, anxiously awaiting the soon-to-be-delivered feast. Shortly thereafter, the waiter returns with the split pea soup.

"Waiter," the customer gasps. "Your finger is in my soup!"

"That's okay," responds the waiter. "It isn't that hot. It doesn't hurt a bit."

"Waiter, there's a fly in my split pea soup."

"Don't complain, sir. Just give me the fly and I'll get you a split pea."

"Waiter, there is a fly in my soup!"

"Not so loud, sir. Everyone else will want one."

"Waiter, there is a fly in my soup!"

"That's very possible, sir. The chef used to be a tailor."

"Waiter, there is a fly in my soup!"

"No problem, sir. I'll get you a fork."

"Waiter, there is a fly in my soup."

"I know, sir. We ran out of fly spray and now we're trying to drown the little fellows."

"Waiter, there is a fly in my soup!"

"Well actually, sir. That isn't a fly at all. You see, our last customer was a witch doctor and he wasn't at all pleased with the chef's cooking."

"Waiter, what is this fly doing in my soup?"

"It looks like the backstroke, sir."

"Waiter, there is a fly in my soup!"

"Oh, that's not unusual, my good man. The chef used to be a fisherman."

"Waiter, call the *maitre d'*! There is a spider in my soup!"

"Don't worry, sir. Just give me the spider and I'll get you a fly."

"Waiter. There is a fly in my bird's nest soup!"

"What does the honorable customer expect for 25¢? An eagle?"

"Waiter, what's this fly doing in my ice cream?"

"I think he (or she) is learning to ski."

"Waiter, what's this fly doing in my alphabet soup?"

"Most likely he's learning to read. Bright little fellow, isn't he?"

"Look now, you. This coffee tastes like mud!"

"Well, it should. It was just ground this morning."
"No! You don't understand. It really tastes like swamp water!"
"Thank you, sir. We'll have to mention that on the menu."
"Waiter, my wife and I went to the Caribbean last summer."
"Oh, really. Jamaica?"
"No, she wanted to go."

Service Industry Feedback — Real Time (Hotel)

A family of four has caught a flight at 6:00 p.m. from Los Angeles heading to the Hawaiian Islands for a week of rest and relaxation. The flight takes about 5 hours, and, because a couple of hours are gained by flying west, they arrive in Honolulu around 9:00 p.m. They get their bags, catch a taxi, and arrive at their hotel by 9:40 p.m. Checking into their room, they find that the television is not functioning. A call is made to the front desk, and a bellman is sent up to move them to another room.

The desk clerk fills out some sort of failure report. This report specifies the location of the problem (room number) and the nature of the failure (television not functioning properly). In addition, the clerk also notes that the room itself is out of order so that the room will not be assigned to anyone else checking in until the problem is corrected. Finally, the clerk forwards the report to the office of the hotel engineer responsible for correcting such problems. The procedure requires that the engineer fixing the problem notify the front desk when done so that the "out of order" designation can be removed and the room made available.

Unsolicited Letters

In addition to questionnaires, surveys, and real-time responses, a key feedback mechanism is unsolicited letters to management. Such letters often are termed "orchid letters" when expounding on the virtues of an employee or program that a customer or guest greatly enjoyed. Management may use these letters to reward outstanding employees. The letters are also taken into consideration when analyzing changes to existing programs.

Of equal if not greater importance are the unsolicited letters from dissatisfied customers. Although most customers will utilize a real-time method of providing negative feedback, putting the negative comment(s) in writing often will facilitate some sort of response from management which can possibly appease the situation. Management has the opportunity to formally document the failure noted by the customer and, as with the faulty hotel television, route the notice to someone in a position to correct the matter.

To be effective, control must always involve feedback of information about the degree of conformance to the quality and safety standards. Feedback can take many forms. On the production line, both manufacturing personnel and quality assurance personnel provide feedback. Perhaps the most valuable source of feedback comes from the customer. Customer feedback plays a major role in maintaining a sound quality assurance system. The technical paper *Full-Cycle Corrective Action for Improved Warranty Service* (Cappels, 1983) gives the following model for efficient use of feedback:

> "Upon receipt of a defective product, the warranty organization originates a Failure Notice (FN). The FN contains the nomenclature of the product, the date the failure is recorded, the serial number and lot number, if applicable, and the nature of the failure. Depending upon the warranty policy of the company, the company service representative may replace the defective product or issue a receipt to the customer until repair is complete. The Failure Notice and the failed product are forwarded to Quality Engineering for review. The failure data are then entered in the database.
>
> "Upon receipt of a Failure Notice and the failed hardware, Quality Engineering first searches the database for historical failure data that may be used in determining the failure mode. Once the appropriate Corrective Action (CA) has been determined, the Corrective Action is input to the database and forwarded to the performing organization(s)."

Summary

Management, staff, and consultants lead the drive towards standardized processes and products and implementation of efficient feedback mechanisms. These quality management systems must be closely reviewed from a financial viewpoint to be sure that the correct actions are taken. The financial education endorsed by FFQ helps to ensure that the full value of quality management systems materializes.

Self-Study/Discussion Questions

1. Has your company benefited, or could your company benefit, from ISO 9000 or 14000 certification? Why or why not?
2. How is it beneficial to have two different computer systems (Macintosh and PC) on the market? How is it nonproductive?

References

Cappels, T.M., Full-cycle corrective action for improved warranty service, in *Proc. of the 1983 Annual Reliability and Maintainability Symposium,* Institute of Electrical and Electronics Engineers, pp. 20–22, 1983.

Elliott, J.E., *The Philosophy of Process Management,* Lockheed Martin Missiles & Space, Sunnyvale, CA, 1992, p. 3.

Harrington, H.J., *Total Improvement Management,* McGraw-Hill, New York, 1995, pp. 142–143.

Roll, R.R. and Cappels, T.M., *Tom and Bob: Songs, Dances, and Snappy Patter,* Beyond the Yellow Brick Road radio program, KSJS Radio, San Jose, CA, 1973.

5 Quality Control

Introduction

The science of quality began with applications to manufacturing processes and initially was intended to reduce failure rates and improve reliability of products. This section serves as an introduction to basic quality terms and concepts. In no way does it attempt to address every lesson involved in the science of quality. In fact, no document could achieve such a goal because pages are being added to this knowledge bank every day. Greater quality control detail is presented in Chapters 6 and 7, which discuss recommended methods for ensuring the quality, reliability, and safety of products.

Dr. W. Edwards Deming

Dr. W. Edwards Deming has been credited with providing much of the foundation for the evolving science of quality management. Deming's *A Theory for Management* offers 14 points for achieving success in the work place:

1. Create constancy of purpose toward improvement of product and service, with the aim of becoming competitive, staying in business, and providing jobs.
2. Adopt the new philosophy. We are in a new economic age. Western management must awaken to the challenge, must learn their responsibilities, and must take on leadership for change.

3. Cease dependence on inspection to achieve quality. Eliminate the need for inspection on a mass basis by building quality into the product in the first place.
4. End the practice of awarding business on the basis of price tag. Instead, minimize total cost. Move toward a single supplier for any one item, establishing a long-term relationship of loyalty and trust.
5. Improve constantly and forever the system of production and service to improve quality and productivity and thus constantly decrease costs.
6. Institute training on the job.
7. Institute leadership (see point 12). The aim of leadership should be to help people and machines and gadgets to do a better job. Leadership of management is in need of overhaul, as is leadership of production workers.
8. Drive out fear, so that everyone may work effectively for the company.
9. Break down barriers between departments. People in research, design, sales, and production must work as a team to foresee problems of production and in use that may be encountered with the product or service.
10. Eliminate slogans, exhortations, and targets for the work force asking for zero defects and new levels of productivity.
11. Eliminate:
    a. Work standards (quotas) on the factory floor; substitute leadership.
    b. Management by objective (i.e., management by numbers or numerical goals); substitute leadership.
12. Remove:
    a. Barriers that rob the hourly worker of his right to pride of workmanship. The responsibility of supervisors must be changed from sheer numbers to quality.
    b. Barriers that rob people in management and in engineering of their right to pride of workmanship. This means abolishment of the annual or merit rating and of management by objective, management by the numbers.
13. Institute a vigorous program of education and self-improvement.
14. Put everybody in the company to work to accomplish the transformation. The transformation is everybody's job.

Deming is also considered to be the father of statistical process control, which is discussed below.

Process Quality Control and Statistical Analysis

Quality control goes far beyond simple inspection and test. The science of quality is sophisticated and ever expanding. Complex sampling techniques and statistical analyses offer several advantages:

1. Reduction in the cost of inspection
2. Reduction in the cost of rejections
3. Maximization of the benefits from quantity production
4. Attainment of uniform quality, even though inspection testing is destructive
5. Reduction in tolerance limits

Process quality control and statistical process control aid in achieving high levels of quality. The text *Process Quality Control* (Ott and Schilling, 1990) addresses this area in terms of the following three key aspects:

1. *Process control:* Maintaining the process on target with respect to centering and spread
2. *Process capability:* Determining the inherent spread of a controlled process for establishing realistic specifications, use for comparative purposes, and so forth
3. *Process change:* Implementing process modifications as a part of process improvement and troubleshooting

Diminishing Returns of Quality Control Measures

Advantages such as those listed above impact a company's profitability. However, while one would expect such an impact to always be positive, Financially Focused Quality (FFQ) recognizes that a point of diminishing returns exists. A financially focused mindset must be instilled in those responsible for administering the science of quality.

Businessmen recognize that it is often beneficial to implement quality measures. As more money is spent on such prevention costs (training assemblers, tightening inspection criteria), there will generally be less

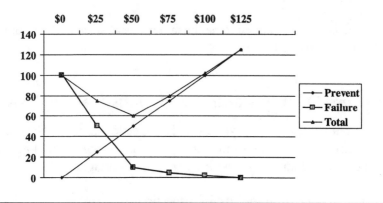

Figure 5.1. Point of Diminishing Returns

money spent on failures. This is in line with the old adage: "An ounce of prevention is worth a pound of cure." The philosophy goes a step further, with formulaic expressions showing that eventually, as prevention costs go up, a point will be reached where failure costs do not decrease at a greater proportion than prevention costs increase. Eventually, a point of diminishing returns is reached.

Figure 5.1 shows that as prevention costs increase, failure costs decrease. The *x*-axis has six points, showing an increasing level of expenditures for preventative costs. When $0 are spent on preventative costs, there are $100 of failure costs, yielding a total cost of $100. When $25 is spent on preventative costs, failure costs drop to only $50, and total costs are $75. The point of diminishing returns is hit when $50 is spent on prevention. Failure costs are only $10, and total cost is $60. When $75 is spent on prevention, failure costs are only $5, but total cost is $80.

A good analogy can be drawn with the value of an automobile tune-up. Most mechanics concur that is it wise to tune up the engine periodically. It is probably better to get a tune up every 50,000 miles than to wait and have it done every 100,000. And it is probably better still to get a tune-up every 30,000 miles as opposed to every 50,000. But, eventually you will reach a point of diminishing returns. Is it practical to tune up a car every 5 miles?

Aside from minimizing costs and increasing profitability, the goal of FFQ is superior quality control. When considering the statement, "Quality is a thought that resides in every executive's mind," one must examine the precepts of FFQ to understand the financial impacts of perceived quality improvements.

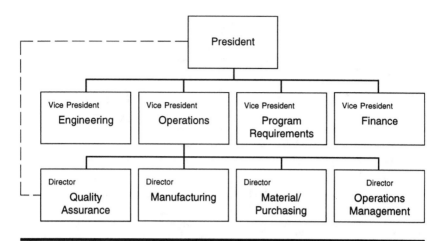

Figure 5.2. Placement of Quality Assurance in Corporate Organization Chart

Quality Control Definition and Goals

The Handbook of Industrial Engineering and Management broadly defines quality control as "those activities which ensure that quality creation is performed in such a manner that the resulting product will in fact perform the intended function." In this sense, Quality Assurance strives to achieve several goals. The first is to ensure that the product characteristics selected will achieve the intended result. A second goal of this function is to ensure that the items produced will be manufactured to the desired quality level. In some organizations, Quality Assurance responsibilities have been expanded to include evaluation of the safety performance in the original design. Thus, a third goal of Quality Assurance is for customer verification of safe performance.

Quality Assurance Organization

The relationships among the many industrial functions within a manufacturing concern must enable synergistic operation. Therefore, responsibilities and lines of communications must be established clearly. Figures 5.2 and 5.3 are examples of organization charts which, respectively, show the placement of the Quality Assurance organization within a company and the various tasks within the Quality Assurance function. The responsibilities of Quality Assurance personnel are varied, product oriented, and tailored to the needs of the individual company.

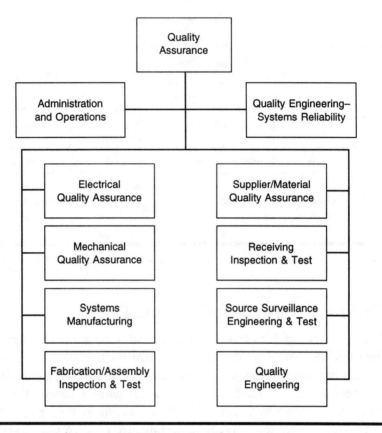

Figure 5.3. Quality Assurance Organization Chart

Notice in Figure 5.2 that there is a dotted line between the President and the Director of Quality Assurance. Many businesses find it advantageous to place the Quality Assurance Director immediately below the President. This figure reflects only one of many possible Quality Assurance placements.

Closed-Loop Corrective Action*

The Fleet Ballistic Missile (FBM) Business Unit within Lockheed Martin Missiles & Space (LMMS) is the prime contractor for the U.S. Navy's

---

* Pages 68 to 76 were previously published in 1994. The material presents the operating environment at the time and is not intended to be a statement of current practices at Lockheed Martin Missiles & Space.

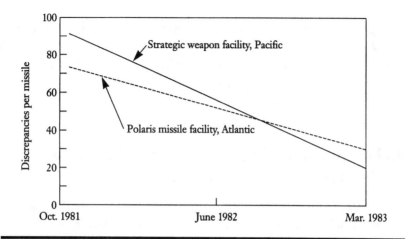

Figure 5.4. Improved Missile Quality

fleet of Trident II missiles. These missiles are critical components of the U.S. defense program. Considering the nature of these products, quality and reliability are of paramount importance.

The Closed-Loop Corrective Action (CLCA) program of the FBM Business Unit maintains an enviable record of quality and reliability. The objectives of CLCA are

1. Document problems and failures of missile components and surface support equipment prior to and after delivery to the government.
2. Verify failures.
3. Diagnose failures, determine cause, and establish the area of responsibility.
4. Recommend and initiate Corrective Actions.
5. Follow up to verify effectiveness of Corrective Actions.
6. Maintain comprehensive statistical records.

A review of such statistical records reveals that the FBM Business Unit has indeed satisfied these objectives. Data were collected to determine the effectiveness of implemented Corrective Actions. As Figure 5.4 shows, there was a steady decline in discrepancies per missile at the Navy's shore-based missile facilities during the first few years of CLCA implementation. High levels of quality have consistently been maintained.

The FBM Business Unit's CLCA system is unique in that it has upper management direction and lower level support to help achieve the safety and reliability goals set by the Navy. The CLCA system is an integral part of the quality and reliability program for one of the United State's largest defense contractors.

Relationship of Financially Focused Quality to Closed-Loop Corrective Action

Of course, CLCA as established by a major aerospace and defense contractor would not be appropriate for most manufacturing concerns. However, features of this system have been incorporated into Financially Focused Quality, which has been successfully applied in various areas within LMMS as well as in other companies and industries.

Closed-Loop Corrective Action, as described here, offers features that guarantee beyond a shadow of a doubt that the highest levels of quality and reliability will be achieved. FFQ, while adopting several of the concepts offered by CLCA, completes the Corrective Action cycle. The merging of a program designed for highest possible quality with tools and techniques to enhance cost-effectiveness offers companies the opportunity to realize maximum quality at a minimum cost. LMMS purchases materials, manufactures products, and then services these products after delivery to the customer. In order to provide the required quality and reliability, CLCA has three facets, each of which interacts directly with the Quality Engineering organization (Figure 5.5). Quality Engineering is responsible for administration of the Closed-Loop Corrective Action.

CLCA Operations and Failure Reporting Systems

Each facet demonstrates the tools and methods utilized by the FBM Quality Engineering Organization.

Supplier Operations

The Supplier Operations organization of the FBM Business Unit encompasses the design and quality requirements placed upon a supplier. For purposes of this section, only the quality requirements will be

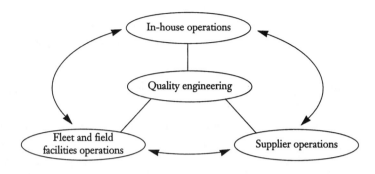

Figure 5.5. Closed-Loop Corrective Action

addressed. A critical part of the contracting process commits the supplier to comply with specific quality system requirements. These requirements are included within the Supplier Failure Reporting System.

The contracts established between the U.S. Navy and the FBM Business Unit require that LMMS manufacture missiles within the constraints of ordnance documents, military specifications, and unique contractor-prepared documents as approved by the Department of the Navy's Strategic Systems Program Office. In turn, the FBM Business Unit then places specific requirements on its suppliers through systems documents, Product Assurance work statements, special application documents, and supplier data requirements lists.

The Supplier Quality System has been established to assist the supplier in performing the functions required in these documents. This system includes:

1. Product Assurance Action Reports, to document Corrective Action
2. Product Assurance Action Reports through Failure Reporting and Failure Diagnosis
3. Vendor Request for Information or Change
4. Test Equipment Downtime Reports, when problems arise from test
5. Equipment Discrepancies

Supplier Operations provides assistance to the suppliers through Product Assurance supplier representatives. Depending upon contractual requirements, such representatives may be provided on a resident basis or on an itinerant basis. The supplier tests its products per contractual requirements. If the hardware fails the test, an analysis is performed to determine which of the following reporting documents should be utilized:

1. *Product Assurance Action Report:* Mandated by additional product assurance requirements being placed upon suppliers of major complex components
2. *Vendor Request for Information or Change:* Required any time a piece of hardware does not meet blueprint requirements but is deemed usable by the acknowledged authority (i.e., Material Review Board)
3. *Test Equipment Downtime Report:* Provides documentation that supplier test equipment is meeting contractual obligations

Once determined, the relevant documents are completed and forwarded to FBM's Quality Engineering group where a Corrective Action (CA) is established.

The primary objective of FBM's Supplier Operations organization in the CLCA system is to monitor and report progress made on such Corrective Actions. This objective is accomplished through the Product Assurance supplier representative. This representative has access to the subcontract and product assurance requirements imposed on the supplier and has insights into and knowledge of the supplier's facility. In addition, Quality Engineering's CLCA Coordinator provides the supplier representatives with copies of pertinent rejection and Corrective Action documentation.

In-House Operations includes Manufacturing, Inspection, and Quality Engineering organizations. The Factory Failure Reporting System is an integral part of CLCA (Figure 5.6). The Inspection Division of the FBM Business Unit initiates most of the failure documentation for In-House Operations within the CLCA system. Inspection performs such major functions as:

1. Receiving inspection
2. Receiving acceptance test
3. Mechanical in-process inspection
4. Electrical in-process inspection
5. Electrical test inspection
6. Systems checkout
7. Shipping inspection

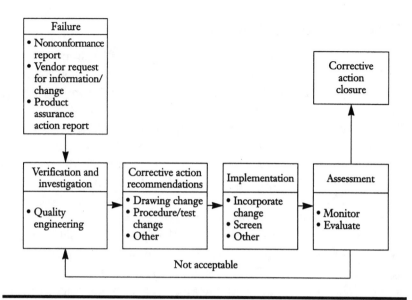

Figure 5.6. Closed-Loop Corrective Action: In-House Failures

Descriptions of these functions are contained later in this chapter (Quality Assurance Functions).

Regardless of which inspection function detects a failure or discrepancy, the proper failure reporting document is initiated. The inspector prepares Discrepancy Reports (DRs) for nonconforming materials. The following procedure is followed when a DR is prepared:

1. Proper entries are made in production paper (e.g., log books, line flows), and notification is given to the responsible organization that a Corrective Action is required. When the CA can be taken immediately, as in the case of correcting an operator error, the CA is documented and the DR is signed. The DR then continues through the routine processing.

2. When the CA cannot be taken immediately, the problem is referred to the Quality Engineering Organization, where a Product Assurance Action Report is initiated and tracked by the CLCA Coordinator. The person identifying the failure prepares Trouble Failure Reports, under certain conditions, on fleet returned hardware.

Fleet and Field Facilities Operations

If an anomaly or test failure occurs after a piece of hardware has been delivered to the Navy, a Trouble Failure Inspection/Rejection Report is written at the shore-based facility or a Trouble Failure Report is written by the Fleet. Because there are many contractors other than LMMS in the Fleet Ballistic Missile system, the Navy designated the Naval Warfare Assessment Center to receive and process all failure documents. Such documents are forwarded from the Navy Warfare Assessment Center to the respective contractors. The FBM Trouble Failure Report Processing Group receives these documents for processing and inclusion into the CLCA System.

When required, a positive Corrective Action response is made to the originator of the failure document in the form of a Trouble Failure Report/Corrective Action Report. The response made directly to the originator of the failure document is part of the reason this process is termed "closed-loop". It closes the loop started by the originator initially documenting the failure. Trouble Failure Reports/Corrective Action Reports are initiated by Quality Engineering pursuant to the CLCA procedure. Formal reports are published each month by the Trouble Failure Processing Group for external and internal distribution. These reports provide the status of open and closed transactions.

Quality Engineering

Quality Engineering is responsible for administration of the Closed-Loop Corrective Action. There are two primary roles falling under the Quality Engineering umbrella:

1.　Quality Engineering Corrective Action Engineer
2.　Quality Engineering Closed-Loop Corrective Action Coordinator

The key individual in the CLCA system is the Corrective Action Engineer. This engineer is responsible for specific types of hardware. Specific hardware assignments are made to individuals with the expertise needed to solve the problem.

When inspection or test detects a failure or discrepancy, the responsible engineer must verify the failure and ensure that the failure is properly documented. When the failure is verified and properly documented, the engineer takes the action necessary to obtain the failed hardware for examination.

Scientific labs are available for analysis of dynamics and theoretical aspects of the failure. The engineer also has access to a Materials and Process Lab for analyzing metals and materials for chemical and physical properties. When diagnosis and analysis have been completed, the engineer pursues a specific Corrective Action. A Corrective Action Product Assurance Action Report is a document that outlines the failure analysis and diagnosis and recommends that corrective action be taken. The engineer initiates this action report document and forwards it to the responsible organization — FBM Design Manufacturing or the supplier.

The responsible organization is required to respond with the following data:

1. Description of the Corrective Action taken
2. Specifics of hardware (serial number, lot number, etc.) and date of the Corrective Action implementation

The engineer later verifies that the Corrective Action has been taken as stated on the Corrective Action Product Assurance Action Report.

The Material Review Board is an organization staffed by Navy and LMMS employees. This board reviews rejection documents to ensure that an adequate corrective action statement has been made. The Corrective Action Engineer is contacted when the Material Review Board questions the validity of such statements. When contacted, the engineer must document the validity or begin the Corrective Action process.

The Closed-Loop Corrective Action Coordinator is the point of contact for all Corrective Action activities. The CLCA Coordinator is notified of failures and discrepancies and tracks the Corrective Action process through completion. Factory and supplier failures are processed directly by the coordinator, as are field and fleet failures. The failure is assigned to the responsible Corrective Action Engineer. With the CLCA Coordinator responsible for nontechnical documentation and status of the Corrective Actions, the Corrective Action Engineer is free to perform the technical aspects of failure diagnosis and corrective action. Corrective Action assignments, estimated completion dates, and interim and final status reports are maintained by the CLCA Coordinator. Reminders are issued via computer for overdue verification of corrective action.

The Data Engineering group provides critical support to Quality Engineering. There are two major functions performed by Data Engineering:

1. Establish and maintain master data files regarding missile hardware failure and discrepancy and Corrective Action documentation.
2. Provide timely failure, discrepancy, and quality trend data to organizations which are requested to take or ensure corrective action.

Summary

It should be obvious, based on the above description, why the FBM Business Unit of Lockheed Martin Missiles & Space has attained an enviable record of excellence in quality and reliability. The nature of the FBM product requires this system for total quality control. CLCA contains virtually every conceivable aspect of a top-notch Corrective Action program. That is why it was used as a model from which Financially Focused Quality was developed.

Quality Assurance and the Product Life Cycle

The responsibilities of the Quality Assurance organization throughout the product life cycle are discussed below.

Research and Development Phase

During the research and development phase, the need for a new product is identified and potential methods for satisfying this need are examined.

Design Phase

The product is completely defined in the design phase, then the design is tested to ensure proper operation. The role of Quality Assurance is critical in this phase. Ideally, Quality Assurance test engineers can identify all design anomalies prior to beginning formal production. Testing is extended to include proper operation of the product in various environments (e.g., light, temperature, salt spray, humidity, vibration, shock, sand, dust, rain, sunshine, etc.) in which the product is expected to operate. The design phase concludes when the product has been completely identified and qualified by test for each environment.

Production Phase

It is in the production phase that the role of quality control is perhaps best understood. The production phase requires a complete Quality Assurance check of the drawings and inspection of hardware to ensure complete compatibility. Any changes in the design should once again undergo full qualification testing, and the product definition should be continuous so that each configuration always can be included in the active drawings and technical data package. The definition and hardware must always match, with all nonconformances being controlled.

Operation Phase

As the product enters the operation phase, it becomes available to the customer. The user must be trained adequately for proper operation. Both the manufacturer and user must work together in developing a maintenance plan and operational system to ensure that the product is utilized properly.

Disposal Phase

This phase begins when a decision has been made that the useful life of the product has ended. Quality Assurance tests are used to make this determination. Nuclear systems have made the public more aware of the disposal phase, as nuclear waste cannot be stored in junkyards like old automobiles. The life cycle summarized above is true for a variety of products: appliances, aircraft, boats, farm equipment, consumer products, trains, ships, nuclear installations, and more. To be effective, each phase must be scheduled and completed in the proper sequence.

Cycle of Quality Activities

Figure 5.7 presents an overview of quality activities and their interdependent relationships. As shown in the figure, the Quality Assurance organization interacts with virtually every group involved with the product. Quality Assurance begins with program control functions in the product planning and development stage. The next step is developing test criteria and design control as the product undergoes design engineering. Quality Assurance material control functions play a major

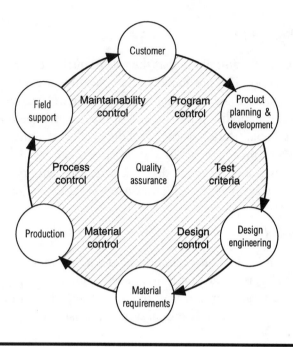

Figure 5.7. Cycle of Quality Activities

role when material requirements are addressed. Along with production, the Quality Assurance function performs inspection, test, and process control. Field service is supported by maintainability control activities.

Figure 5.7 illustrates the flow of activities from the development of a product through the requisite engineering and material decisions which lead to the production of the product that eventually reaches the customer by way of field service. The cycle begins again when customer comments and evaluations are used in the next product planning and development venture.

Quality Assurance Functions

An overview of Quality Assurance functions and definitions follows.

Receiving Inspection

Receiving inspection of components to be used in the manufacturing process is usually performed on receiving docks and receiving areas throughout the company. Such inspection includes simple identification and inspection for damage during shipment. An inspection technique known as sampling is used when many items are received at the same time. The size of the sample lot to be inspected varies. As more parts are found to be out of spec (not compliant with required specifications), the quantity sampled increases.

In-Process Inspection

As the manufacturing process progresses, inspectors are in place to ensure that the product is coming together as designed. In government contracting, the government has been known to want the inspector to perform no manufacturing and the manufacturer to perform none of the inspection. This is to establish a system of checks and balances in the process; however, this can be very costly. Often, the inspector will wait for the manufacturing worker to perform a function, and then the manufacturer waits while the inspector uses tools to inspect the work. Some government contractors have significantly reduced costs by developing a "production partnership" which involves, among other things, training manufacturers in the skill of inspection, thus allowing manufacturing employees to inspect their own work

Shipping Inspection and/or Final Inspection

Shipping inspection and/or final inspection includes examination and testing of products prior to shipment to customers to determine whether they conform to specifications. Any final tests are also performed at this stage to make sure the customer is getting what is wanted.

Accelerated Test

An accelerated test involves testing at abnormally high stress or environmental levels in order to induce earlier failures. Extrapolation is used to convert the short life under severe conditions into expected life under normal conditions.

Configuration Management

Configuration management is a discipline applying technical and administrative direction and surveillance to identify and document the functional and physical characteristics of a configuration item, to control changes to these characteristics, and to record and report change processing and implementation status.

Destructive Testing

Destructive testing is that testing that may impair the subsequent usefulness of a product.

Engineering Evaluation Test

Engineering evaluation tests prove that items meet function environmental design limits.

Electrical In-Process Inspection

Electrical in-process inspection of electrical hardware is performed during the manufacturing or repair cycle in an effort to prevent defects from occurring and to inspect characteristics and attributes which cannot be inspected at final inspection.

Fabrication and Assembly Inspection and Test

Fabrication and assembly inspection and testing are performed during the manufacturing or repair cycle in an effort to prevent defects from occurring and to inspect the characteristics and attributes which cannot be inspected at final inspection.

Failure Analysis

Failure analysis is a series of actions performed on a verified failure to separate the problem into parts or elements for examination and determination of specific cause of failure and corrective action required.

Failure Diagnosis

Failure diagnosis is a component of failure analysis and is a planned physical examination to determine cause of failure.

Failure Verification

Failure verification is a series of actions taken to determine whether or not the hardware is discrepant or if a problem is traceable to operator error, test equipment, or procedural problems.

First Article Compatibility Test

The first article compatibility test is testing of the first unit produced in the first production run to confirm that the item meets drawing requirements.

Material Review Board

The Material Review Board is a formal contractor/government board established for the purpose of reviewing, evaluating, and disposing of specific nonconforming supplies or services and for ensuring the initiation and accomplishment of corrective action to preclude recurrence.

Mechanical In-Process Inspection

Inspection of mechanical hardware is performed during the manufacturing or repair cycle in an effort to prevent defects from occurring. It is also performed to inspect the characteristics and attributes that cannot be inspected at final inspection.

Metrology Audits

Metrology audits are the systematic review and evaluation by technical specialists to determine the adequacy of the contractor's system for calibration and measurements.

Parts Application Review

Parts application review is an engineering analysis of the effects of electrical, environmental, operational, and packaging stresses on part reliability.

Product Evaluation Test

A product evaluation test evaluates factory screening and design margins.

Quality Assessment

Quality assessment is a critical appraisal that includes qualitative judgments about an item (e.g., importance of analysis results, design criticality, and failure effect).

Quality Audit

A quality audit is an independent review conducted to compare some aspect of quality performance with a standard for that performance.

Receiving Inspection and Test

Receiving inspection and test include the examination and testing of purchased materials to determine whether they conform to specifications.

Source Acceptance

Source acceptance is the validation of supplier acceptance of hardware in accordance with acceptance criteria; the test is performed at the supplier's facility.

Source Verification Inspection

Source verification inspection takes place at the supplier's facility to determine conformance to the specified process, inspection, and test requirements.

Statistical Process Control

Statistical process control is a program for ensuring that the product of a manufacturing or inspection process meets the requirements of the engineering specification.

Value Engineering

An organized effort is directed at identifying the functions of a product, either hardware or software, in order to achieve the required functions at the lowest overall cost consistent with the requirements of performance, reliability, quality, and maintainability.

Summary

These functions are a sampling of those that are performed in manufacturing enterprises. It would not be practical to list all of the Quality Assurance functions, but many other such functions in support of most activities required to deliver products to the customer could conceivably exist.

Traditional Finance Involvement in the Cycle of Quality Activities

Figure 5.8 depicts the relationship of the Finance function to quality activities in an environment where Financially Focused Quality is not used. Finance traditionally participates as an outsider, reacting to cost and schedule data involving many events — sales to customers and returns from customers result in debits and credits to the income ledger.

Product development and planning activities cause the Accounts Payable department to process checks for market research consultants and the administration of miscellaneous costs associated with questionnaires, surveys, focus groups, research and development, and bids and proposals for new products. Other costs for which there must be accounting and payment include:

1. Test and analysis of engineering designs
2. Subcontractor review and audit
3. Inspection and analysis of materials used in the manufacturing process
4. Inspection and test of in-process assemblies and subassemblies
5. Final test and checkout of product before shipment
6. Review and ongoing analysis of the shipping process

Field support activity requires the Finance department to administer costs associated with:

1. Verifying that the product is installed properly
2. Ensuring that instructions for use and care are clearly presented
3. Providing warranty service as required

The Finance organization is the first to receive reports of such cost data and reacts to such reports by performing many functions, including:

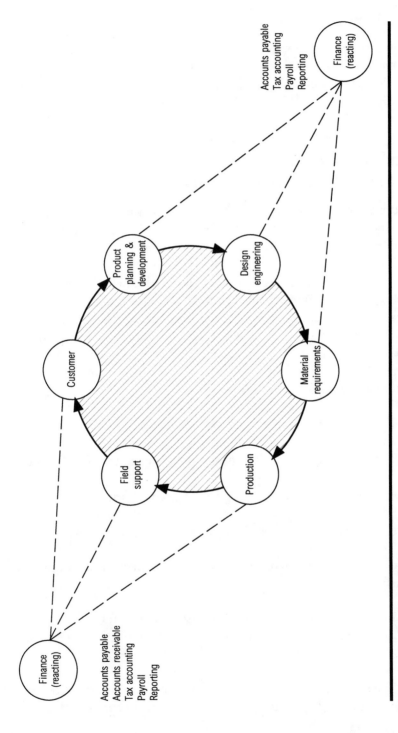

Figure 5.8. Traditional Financial Involvement in the Cycle of Quality Activities

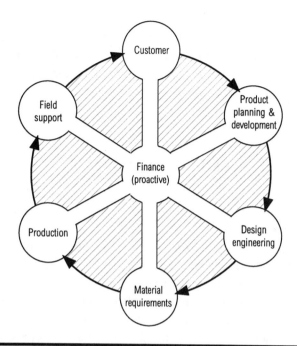

Figure 5.9. Completing the Cycle of Quality Activities: Finance

1. Summarizing
2. Preparing financial reports to management
3. Preparing financial reports for federal, state, city, and local taxing authorities
4. Preparing reports for customers
5. Preparing reports for stockholders
6. Preparing reports for public record
7. Administering accounts receivable
8. Administering accounts payable

The tasks described in this section are critical to the successful financial operation of a company; yet, FFQ provides a means whereby increased financial success can be obtained.

Completing the Cycle of Quality Activities: Finance

Figure 5.9 depicts an overview for successful FFQ implementation. The finance organization is no longer on the outside looking in and merely reacting to the cycle of quality activities. Finance becomes the hub for all business operations, a predominant component of the product delivery wheel around which the other product delivery systems revolve. Future chapters of this text discuss the spokes linking the financial organization to each vital system.

Self-Study/Discussion Questions

1. How are the Quality Assurance measures used in a manufacturing concern similar to those for a service organization? How do they differ?

2. How often do you have your car tuned up or your oil changed? How do you determine the frequency?

3. How important is it, in your company, that its products be 100% error free? What quality control measures are in place? Are more needed? Are there too many? How does your company determine the right balance? Is statistical process control used in your company? What statistical process control processes might be applicable?

References

Cound, D.M. and McDermott, T.C., *The Handbook of Industrial Engineering and Management*, Prentice Hall, Englewood Cliffs, NJ, 1971.

Dewar, D.L., *The Quality Circle Guide to Participation Management*, Prentice-Hall, Englewood Cliffs, NJ, 1982.

Kozich, S., Testing for product liability, in *Transactions of 1978 American Society for Quality Control Technical Conference*, American Society for Quality Control, Milwaukee, WI, 1978.

Ott, E.R. and Schilling, E.G., *Process Quality Control*, Quality Press, Milwaukee, WI, 1990, p. 172.

6   Avoiding Product Liability: Design and Development

Introduction

Financially Focused Quality requires that every conceivable cost impact be considered when taking corrective actions or implementing process improvements. To facilitate the Financially Focused Quality mindset, manufacturing management needs to be aware of existing quality control tools for verifying product safety. As the number of litigations continues to increase throughout the world, the importance of recognizing potential cost impacts resulting from product safety hazards must be emphasized. *Products Liability — Design and Manufacturing Defects* is a very comprehensive text which thoroughly addresses product liability issues, some of which are summarized in this text and reprinted with permission of the publisher.

Product Safety

Product safety is perhaps the most important product characteristic. Thus, product safety may be considered an important characteristic to be "assured" within the scope of a Quality Assurance program. In practice, however, a manufacturer may be justified in treating product safety as a separate subject because of its effect on human well-being, not to mention liability and other considerations.

Responsibility for Product Safety

When a consumer experiences harm or material damage resulting from a purchased product, liability is a key consideration in the ensuing litigation. For example, studies have been performed to determine responsibility for television fires. The televisions in question had undergone tests implying that Underwriters Laboratories, Inc. had verified compliance to specific quality and safety standards. Despite these tests, television fires occurred (Berman, 1977).

Manufacturer Liability

While outside agencies such as Underwriters Laboratories, Inc., can provide valuable services, the case above illustrates that it is incumbent upon the manufacturer to assure the quality and safety of its products. Louis Bass, P.E., J.D., has stated that safety must always be a major consideration in the corporate decision to design, manufacture, distribute, and sell a product. Executive management must be aware that the company can be exposed to product liability if a product it manufactures or sells causes personal injury or property damage to buyers or third parties. A manager involved in the purchase of components and equipment manufactured by others should be aware that contractual provisions in sales agreements may be making the company vulnerable to unnecessary product liability exposure or limitations on the company's remedies. Even with protective contractual language in place, a manufacturer must be ever vigilant for potential failings of suppliers. Such failings may cause harm to their customers and thus their own reputation.

Reputation is an important component of goodwill, often considered by analysts attempting to evaluate the worth of an enterprise. Financially Focused Quality encourages corporate leaders to continually assess the impacts that management actions might have on goodwill.

In Bass' words (Cappels and Bass, 1986): "Product liability was and continues to be the creature of social policy. Considerations of a manufacturer's responsibilities have changed as society's value system has changed." Financially Focused Quality recognizes that, as society's value system continuously changes, manufacturing management needs to attain a clear understanding of product liability. The next section presents a summary of the basic components of a product liability case.

Basic Conditions of a Product Liability Case

A number of different elements must be analyzed as components of a product liability case. In order for liability to exist, each of the following conditions must be proven to exist:

1. The manufacturer must have a responsibility to the plaintiff. In a product liability case, this requires that (a) the defendant must have had an opportunity to foresee that a plaintiff could bring suit; (b) the item must actually be a product; (c) the defendant was in some way involved in the design, manufacture, or distribution of the product; and (d) the process entered the stream of commerce. Bass notes that there may still be liability other than product liability, without these conditions being met.
2. There must be a defect (failure) in the product. For a defect to exist, it must be proved that the product presented an unacceptable risk of injury when the benefits of having the product on the market are weighed against the risks. This portion of the analysis evaluates the state of the art, custom and practice, government standards, and technological and economic design alternatives. The question here is, could the product have been designed or manufactured to reduce the risk consistent with product function, utility, price, and technological feasibility?
3. The product must be the proximate cause of the plaintiff's injury. This analysis evaluates the way in which the product was used or misused and the actions of the plaintiff and others to determine if the product was the legal cause of the plaintiff's injury.
4. The plaintiff must have suffered personal injury or property damage.

Management must be aware of the above conditions when making decisions that could potentially result in a product liability case. Even though many companies have their own in-house legal departments, it is neither possible nor practical to run every corrective action and process improvement by these authorities. It is incumbent upon all levels of management to exercise the Financially Focused Cycle mindset — their own logical faculties — and consider the likelihood of injuries to customers.

Manufacturing Defects

Financially Focused Quality requires that management be constantly prepared to defend their manufacturing processes. This defense should be in regard to not only productivity and direct profitability but also the safe performance of its products. It has been said that litigation involving a manufacturing defect demonstrates the differences between liability under strict liability and negligence. The manufacturer defending its conduct under negligence principles will try to show that it acted reasonably in the manufacture of the product. The manufacturer will argue the reasonableness of the machines and processes used to manufacture the product. The manufacturer will also attempt to show the reasonableness of quality control efforts to identify products with manufacturing defects. Even when a defective product has caused the plaintiff's injury, the manufacturer who is found to have acted reasonably might not be found negligent (Cappels and Bass, 1986).

Quality Assurance vs. Product Liability

Over the past 20 or more years, companies that manufacture and/or sell products have been barraged with litigation involving their product. During this period, some management groups have realized that it takes 20 to 25 times more in sales to pay for the cost of a litigation or claim. This does not include the high internal costs that companies have been experiencing to service their own litigation staff (Jacobs, 1983).

Many manufacturing concerns are recognizing that one of the newest areas of work for an ever-increasing number of quality/reliability specialists involves product safety and product reliability prevention tasks; however, the concepts of Financially Focused Quality do not endorse the automatic hiring of such professionals by a large manufacturing concern. The total financial impact of such actions must be taken into account. The need for product reliability specialists may be better addressed by using part-time consultants. One cannot ignore that every manufacturer needs an individual, consultant, or group of employees to address quality and reliability issues categorized within the function of Quality Assurance.

Evolution of Quality Assurance

The Handbook and Standard for Manufacturing Safer Consumer Products
defines Quality Assurance as being "actions taken throughout the manufacturing process to prevent and detect product deficiencies and product safety hazards" (U.S. Consumer Product Safety Commission, 1975).
The function known throughout the world as Quality Assurance needs
to reappraise its position and fully accept the responsibility for preventing harm to consumers. Liability prevention is no longer a matter for
management to appease by delegation to attorneys in the legal department. Management has been learning that proper attention to basic
Quality Assurance areas can preclude many safety-critical situations,
both in the factory and with the consumer.

Increased emphasis has been placed on safety-related Quality Assurance. The increasing importance of Quality Assurance has been strongly
influenced by the following basic industrial trends and social develop-
ments:

1. The number of litigations resulting from property damage and
    injuries in the field is growing.
2. There has been greater specialization of labor, as the knowledge
    base continues to expand.
3. The precision and complexity of products are increasing.
4. There is growing consumer awareness of product quality and
    related safety factors which is attributed to radio, television, and
    various publications.

Upper-level management is recognizing the economic hazards of marginal quality, which can result in defects and potential personal injury
and property damage. All managers should recognize that the longevity
of an industrial enterprise today is strongly dependent on two key factors:

1. The nature of the product's quality and safety must be thor-
    oughly understood.
2. The methods employed in achieving the company's quality/
    safety objectives must be comprehensive and viable.

Quality Assurance management is often delegated full responsibility for
ensuring that the manufacturing process produces products of consistent

quality, but of equal if not greater importance is the assurance that the process creates a product free of exposure to product liability issues.

Relationship Between Quality Assurance and Product Liability

Members of the legal profession may view the relationship between the Quality Assurance functions and product liability differently than do representatives of the quality control world. Just as attorneys are not trained specifically in the science of quality, quality professionals are not trained in the details of the different pleadings. Such pleadings include strict liability in tort, negligence, express warranty, or implied warranty. Additionally, the quality profession is not necessarily concerned with legal ramifications, as his job is to prevent defects of product reaching the customer.

Defects have been described in many legislative acts and administrative rulings, many of which are obscure but have valid enforcement proceedings available, which is also an another area of concern. The quality professional is neither schooled in nor expected to be concerned with such legal details as the choice of venue, the differences between various jurisdictions, or why certain rules of evidence are applicable in the different courts. However, he *is* concerned about responding to interrogatories and being deposed.

It is impractical to provide thorough legal training to members of the Quality Assurance profession. At the same time, a company cannot be expected to educate a member of the legal profession in the complete science of quality control. A valid course of action would be to adopt a systematic approach — from executive management down — to quality and safety assurance.

Systematic Approach to Quality and Safety Assurance

Safety-related quality objectives are more easily attained through developing a systematic approach to the control of quality. A systematic approach to quality and safety should include the functions addressed in the text below (Cound and McDermott, 1971).

Product Design Safety

There are five major activities in the systematic approach to product design safety.

1. Provide assistance in evaluating customer requirements to ensure clear understanding of product safety objectives.
2. Review design documentation for conformance to design standards and practices and for identification of potential safety problems. This should include submittal to the required safety agencies and remedial action when indicated.
3. Validate completeness and accuracy of qualification tests and design proof tests.
4. Audit release and distribution of design documentation to ensure that drawings and specifications are current and accurate.
5. Provide information on previous safety-related problems for consideration in new product designs or current product improvements.

Purchased Material Quality

There are four key components relating to purchased material quality:

1. Assist in evaluating potential subcontractors or suppliers.
2. Review subcontracts and purchase orders for completeness and correctness of safety-related quality requirements.
3. Ensure that purchased material meets purchase order requirements.
4. Initiate Financially Focused Quality with subcontractors and suppliers when purchased material is not of acceptable quality level.

Manufactured Product Quality

Procedures should be established to achieve the following nine controls:

1. Evaluate and approve manufacturing equipment, processes, testing, and test equipment.

2. Ensure that measuring and test equipment is properly calibrated and maintained.
3. Establish inspection points and develop methods and instructions.
4. Perform inspections and tests at selected points in the manufacturing process.
5. Collect and analyze inspection and test data.
6. Provide information on process and product quality levels.
7. Initiate FFQ on out-of-control conditions and related quality problems.
8. Conduct follow-up via the Process Improvement Follow-Up Plan (see Chapter 8) to ensure timely accomplishment of cost-effective corrective action.
9. Control handling, preservation, and packaging of material and equipment from receipt through shipment of final product. Within this activity are steps to ensure that materials are handled and stored properly and that the history of the material is documented.

Product Safety After Delivery

Equally important as design, quality of purchased parts, and manufacturing safety is the safety of the product in the hands of the ultimate user. Toward this goal, there are five important controls:

1. Ensure that product service publications are clear and correct.
2. Verify that spare parts conform to safety requirements.
3. Ensure that company-performed repairs and modifications are administered in accordance with company product safety requirements.
4. Gather and analyze complaint or accident data from the field to measure the degree of product safety in use and to initiate appropriate FFQ activity.
5. Put a follow-up mechanism (e.g., Process Improvement Follow-Up Plan) in place in case problems emerge in the field.

The Financially Focused Quality approach to quality management is not content with assigning prime responsibility for quality to the Quality Assurance group. Such responsibility rests equally with quality and finance. The financial involvement results when all employees consider the financial impacts of their activities. Averting product liability is a significant area in which profitability can be enhanced.

Quality Control and Product Liability Aversion

Most manufacturing concerns wishing to develop a well-run risk aversion program should perform the following two steps (Geoffrion, 1978):

1. Management must first develop a clear and unequivocal policy which basically states that the company's goal is to provide its customers with safe products
2. Management must then establish a mechanism to accomplish the stated policy. This mechanism could take the form of a multi-disciplined committee charged with product safety and liability exposure considerations.

The committee may be made up of members of management who meet periodically to discuss product safety and liability issues. Many of these issues may surface via submittals of process improvement recommendations (see Chapter 8). In these cases, the Corrective Action Coordinator would present the concerns whenever the committee met. Also, to reduce product liability risk, a manufacturer must ensure efficient acquisition and utilization of product safety-related quality information, as discussed below.

Product Safety-Related Quality Information

Quality Assurance is strongly dependent upon the acquisition and utilization of information to perform its function effectively. Required product safety-related quality information generally falls into three basic categories (Cound and McDermott, 1971): (1) status information, which includes records and reports providing visibility and status of actions occurring during the product flow through the production process; (2) historical information, which consists of records and reports providing objective evidence of actions that can be used to support future problem investigation; and (3) action information, which consists of records and reports used as primary tools in operational and management decisions.

The Financially Focused Approach to Product Safety-Related Corrective Action

Corrective action provides a basis for correcting unsafe or unsatisfactory conditions. Such action is usually one of the most meaningful activities of the Quality Assurance organization. Effective corrective action includes the following steps:

1. Identify the problem.
2. Determine the basic cause of the problem.
3. Generate methods to correct the conditions or factors causing the problem; critical in this step is determination of costs associated with new procedures and processes.
4. Identify and apply best method based on cost effectiveness and safety of the resulting product.
5. Establish a Process Improvement Follow-Up Plan.
6. Carry out the corrective action and follow up as planned.

Product Design: The Beginning

The Quality Assurance function is involved with the product from the very beginning, when the product is initially designed.

Product Design Safety

A respect for quality and for the customer must propagate from executive management to establish an atmosphere conducive to the design of safe, high-quality products. The basic responsibility for the safety of the product design generally falls on the shoulders of the engineering organization. Engineering establishes safety objectives, selects the specific characteristics to achieve these objectives, and develops and publishes the documentation which defines the characteristics.

A clear understanding of the customer's need is a prerequisite to the development and design of products with specific quality and safety characteristics. Product design and development formally start with identification of the customer need and a general concept of how to satisfy it. The Quality Assurance organization is frequently assigned the task of comparing customer detail specification requirements with internal process specifications and practices to determine differences.

In producing consumer goods, the manufacturer frequently faces the problem of determining by whom and how the product will be used. Market research techniques provide a valuable aid in measuring the needs of the customer and his probable reaction to new products. Market research techniques include consideration of competitive product experience and privately held consumer surveys (feedback is discussed in more detail in Chapter 4).

Product Safety Specification

An element vital to the safety of a product design is the safety specification. As stated earlier, usually the engineering organization specifies "product safety" by enumerating in detail the characteristics of the material to be used, the processes to be employed, and the physical and functional characteristics that must be present in the completed item. Such detail is contained in the design documents, including drawings, specifications, and standards.

The engineering organization is assisted by many parts of the industrial team in developing these design documents. To ensure that all segments of the company have a voice in formulating the design documentation, most companies have established a standards committee. The committee allows for the formal coordination of these documents prior to release.

The Quality Assurance organization ensures compliance with design documentation requirements. When required, changes to or temporary deviations from design documentation are authorized by the engineering organization utilizing a controlled change mechanism. Usually, such changes are accompanied by a temporary tightening of inspection criteria or, when a policy of sample inspection is in place, an increase in the frequency of inspections. Within the confines of Financially Focused Quality, the Process Improvement Follow-Up Plan would ensure that the frequency of inspections would eventually decrease as allowed.

Standards for Design and Manufacture

In the context of product liability prevention, a standard has been defined as being a fixed, customary, or official measure (such as a quantity, quality, or price). The key is that the standard be an agreement among the parties involved in its use (Jacobs and Mihalasky, 1976).

Standards are applied to many industrial and business areas such as design, procurement, manufacturing, marketing and distribution, use of the product or service, and service and repair. Standards may cover a very broad range of product specifications, including:

1. Size and strength characteristics
2. Performance in terms of quality, reliability, and safety
3. Assembly methods
4. Inspection, test, and other control procedures
5. Packing and packaging
6. Distribution
7. Identification and documentation
8. Published data for safe use, including warnings, labels, and instructions

These standards often are evaluated to determine the safety of a product design.

Evaluation of Safe Product Design Quality

The selected safety characteristics should be evaluated from an analytical point of view and from actual performance demonstrations. This more often than not involves reviews and tests at various stages of product development. Responsibility for this evaluation varies widely. Usually this evaluation is headed by the engineering organization with assistance from other organizations. In some companies, Quality Assurance or another department independent of engineering may be assigned this function.

Design and Documentation Reviews

Design reviews aid in product development activities. They serve to guide and control the quality of a product so that it meets the needs and demands of the market (Jacobs, 1979). The collective expert knowledge and expertise within a company are utilized during the review in order to ensure that intended qualities are incorporated in a product during the design phase. The reviews also provide a milestone at which some set of criteria must be satisfied before the project can continue toward completion.

Documentation review may be performed as part of the design review or the safety design review, or as a subsequent review to final release of the documentation. The documentation review ensures that the product documentation conforms to the design standards of the company and, at the same time, clearly conveys the intent of the original design.

The design and documentation review teams are comprised of experts representing virtually every organization within the company. The functioning of the team is similar to the operation of the safety design review (see below).

Safety Design Review

In the course of the safety design review, Quality Assurance plays a major role in minimizing or preventing product liability. This is one of the techniques used to evaluate the safety aspects of the product design. The purpose of the review is to provide the designer with a constructive evaluation and suggestions for improving the design.

A team of specialists who are knowledgeable in specialized technical fields or related activities that may have an important bearing on the design performs this review. The nature of the design influences the team composition; however, the typical team members may include:

1. Stress analysts
2. Circuit analysts
3. Safety specialists
4. Safety engineers
5. Quality engineers
6. Production engineers
7. Maintenance engineers
8. Regulatory engineers

The team reviews the following:

1. Details of customer specifications
2. Existing machine and process capabilities
3. Potential safety problems
4. Potential quality problems
5. Safety problems encountered previously that may reoccur

6. Quality problems encountered previously that may reoccur
7. Test results
8. Product design and test documents

As part of the safety design review, participants discuss critical safety characteristics, as discussed in the next section.

Classification of Safety Critical Characteristics

Variations in product characteristics will affect end item safety to varying degrees. As such, classifying characteristics with regard to their relative criticality is advantageous. Through classification, both Quality Assurance planning and operations can be adjusted to provide a higher confidence level regarding attainment of the characteristics which are more critical to end item safety.

Either the organization that established the original requirements or the safety organization should perform the classification. Data gathered from market research techniques should be examined very carefully. It might be determined that additional market research is required. For example, if market research revealed that there was a potential for injury with a product, further data obtained using different research techniques may be required. Design and qualification tests are used to determine if products meet required safety criteria.

Design and Qualification Test

Design and qualification tests are used to determine if products meet required safety criteria. The testing usually include functional and environmental tests (shock, temperature, humidity, etc.), as well as various reliability life tests. The engineering organization in most companies will fabricate engineering models to be utilized in these tests. The "as tested" configuration must be clearly defined and documented. The Quality Assurance organization usually has responsibility for the testing of these models and submission of data and models to governmental or independent safety agencies.

The test equipment should have the required characteristics and must be calibrated. The tests should be performed in a specified manner, and the test results must be accurately documented. The integrity of the engineer conducting the test is also important. His credibility with safety agencies will help smooth the way. It is also not wise to rely on results of tests performed by an engineer with a tainted reputation, lest his testimony not be credible in the event of difficulty later on. Having a safe design, the manufacturer must consider the source of its materials, as discussed in the next section.

Vendor Materials Control

According to Hayes (1982), "Materials control is the process that manages the organization, control, and disposition of all materials from the time that a requisition for materials is issued by production until finished articles are shipped." Vendor materials control is comprised of the systems utilized to ensure the reliability of products obtained from a supplier. This section examines systems designed to assure the quality of raw materials, parts, and components.

Sources for Raw Materials, Parts, and Components

The sources for materials must be determined after the specific characteristics for a safe product have been established. In many cases, industrial enterprises are dependent upon suppliers for these items. The manufacturer's first decision regarding each component is whether to make or buy the particular piece.

Make-or-Buy Committees

Many organizations have the ability to make, as well as buy, a particular item. Thus, it is first necessary to establish which items will be manufactured in-house. The make-or-buy committee is made up of various interested functional groups within the company. Financially Focused Quality requires that the financial interest be represented during the make-or-buy decision. Frequently that financial viewpoint would lean towards a decision to make the product in-house. When a company makes the item, all labor and nonlabor costs enter the allocation base for associated overheads, which results in a lowering of related overhead rates.

Quality Assurance participation in this committee is primarily concerned with an evaluation of quality and safety factors related to the make-or-buy decision. When considering the make-or-buy decision, Quality Assurance should examine safety and quality histories of the manufacturer for each particular type of item. At the same time, consideration should be given to the safety and quality histories of the suppliers of each particular type of item. Quality Assurance should also study the safety and quality problems that might be anticipated for a supplier that might not be experienced with the specialized skills or equipment of the buyer's organization. When it has been determined that an item will be procured, the next step is selecting a supplier.

Supplier Selection

The organization assigned the purchasing function usually has several functions and responsibilities. It must acquire items that conform to the requirements specified and obtain these items in time to support manufacturing schedules, and the items must be procured at a minimum cost.

Ordering Cost-Effective Quantities

Procuring the items at a minimum cost is a task that should be carefully monitored by management. There have been examples where buyers have made purchases far above required quantities in order to achieve the lowest unit cost. This is frequently counterproductive, even though the economies of scale involved in delivering products to customers allow suppliers to sell large quantities of a product at a lower unit cost than low quantities. This is one reason why many retail warehouses are prospering.

The cost per ounce for mayonnaise purchased in a 10-gallon drum is significantly lower than the cost per ounce in a 16-oz container. A Financially Focused Quality mindset would encourage the purchase of a 10-gallon drum if, and only if, the purchaser has performed a cost impact study proving that the 10-gallon drum really is a more cost-effective purchase. This cost impact should reflect how much mayonnaise is used before it goes bad, as well as factors related to convenience, refrigerator space, etc.

Conforming to Requirements

Material, parts, and components directly influence the characteristics of the final product. As a result, it is necessary to establish controls, which provide reasonable assurance that the required characteristics are present in the purchased items. One of the roles of the Quality Assurance organization is to provide this assurance. Documentation of processes, material flows, and histories is just as important for the material vendor as it is for the manufacturer of the product.

The manufacturer must select suppliers who have the capability of furnishing the required raw materials, parts, and components at the proper times and at an acceptable cost. Quality Assurance determines whether the supplier can furnish items which meet the requirements of design documentation.

Three procedures assist in determining a potential supplier's capabilities (Cound and McDermott, 1971). The first is consideration of the supplier's performance histories. The second is a supplier quality evaluation survey, and more valuable data can be gathered by completion of a supplier evaluation questionnaire. Each of these procedures is discussed below.

Suppliers' Performance Histories

If the supplier has previously sold items to the manufacturer, a review of the supplier's quality and safety histories may provide the basis for determining the adequacy of the supplier's quality system. This method is good because the history is a record of demonstrated capability rather than potential capability. However, if the history is not accurately recorded or is not current or if the product to be purchased is significantly different from products previously manufactured by the supplier, the histories may not be applicable.

Supplier Quality Evaluation Survey

There are four conditions that lend themselves to on-site evaluation analysis:

1. If the potential purchase agreement includes items that are sensitive to quality degradation because of storage and handling conditions, then an on-site evaluation analysis would be in order.
2. Such an analysis might also be logical if the supplier has a performance history that is incomplete, inconclusive, or below acceptable quality levels.

3. A third reason for conducting this analysis exists when items cannot be evaluated by receiving inspection without disassembly or destructive testing.
4. The fourth condition pertains to safety-critical items of equipment.

Although they are expensive and time consuming, the supplier survey provides valuable information for determining the adequacy of the supplier's quality system. Standard criteria and associated checklists should be used to make the survey as objective as possible.

Supplier Evaluation Questionnaire

A supplier evaluation questionnaire can provide information to assist in arriving at a decision regarding the acceptability of a supplier's quality system. The questionnaire covers three groups of information:

1. The first group encompasses supplemental historical information about the supplier.
2. The second set of questions is used to assist in planning or making a decision to conduct an on-site quality evaluation survey.
3. The third group contains data used to make a decision regarding the adequacy of a supplier's quality system when the articles involved can be evaluated by receiving inspection without extensive disassembly or destructive testing.

Purchasing Provisions and Specifications

Once the supplier has been selected, the company should ensure that Quality Assurance provisions applicable to the product are imposed upon the supplier in the purchase order. Such provisions may include maintenance of the supplier's quality system, first article or first piece inspection materials certification and test reports, and source inspection. Source inspection provisions in the purchase order may include:

1. In-process inspection by source inspection
2. Acceptance test witnessed by source inspection
3. Final inspection and test by source inspection
4. Government source inspection on certain military contracts

Of course, the contract or purchase order must be mutually acceptable, because this document represents the formal and legal agreement between the supplier and the prime contractor.

The provisions listed above may inadvertently be overlooked in the absence of a systematic method of checking procurement documents. Many potential problems may be avoided by checking early in the procurement cycle.

After a contract or purchase order is given to a supplier, there must be continued communication between the manufacturer and supplier. If the purchase involves safety-critical items, it may be desirable to have periodic conferences with the supplier to maintain a clear understanding of all requirements and to resolve problems of engineering interpretation. As the supplier begins work on a contract or purchase order, a review of his drawings and first fabricated items may bring to light areas of misunderstanding. It is much better to resolve these problems early than to be subject to product liability litigation (Cound and McDermott, 1971).

Summary

This chapter has presented numerous Quality Assurance measures aimed at reducing the risk of product liability litigation resulting from the design and development of a product. The next chapter picks up where this chapter ends by following the product through manufacturing and into the hands of the ultimate user.

Self-Study/Discussion Questions

1. In what ways does consideration of potential product liability issues add to the costs of a company? How can such consideration decrease costs?

2. What products do you know of that originally had unsafe designs?

3. Describe any instances you know of when someone was injured or suffered property damage because of a poorly designed product. What actually caused the injury or damage? How could the consumer have prevented it? How could the manufacturer or distributor have averted the trouble?

References

Berman, G.A., Color television fires: a persistent problem, in *Proc. of the 1977 Reliability and Maintainability Symposium*, West Group, St. Paul, MN, p. 27.

Cappels, T.M. and Bass, L., *Products Liability — Design and Manufacturing Defects*, West Group, St. Paul, MN, 1986, pp. 14, 256.

Cound, D.M. and McDermott, T.C., *The Handbook of Industrial Engineering and Management*, Prentice Hall, Englewood Cliffs, NJ, 1971, pp. 704–705, 725–727.

Geoffrion, L.P., Contribution of quality control to product liability aversion, in *Proc. of the 1978 American Society for Quality Control Conference*, American Society for Quality Control, Milwaukee, WI, pp. 470–474.

Hayes, G.E., *Quality Assurance: Management and Technology*, Charger Productions, Capistrano Beach, CA, 1982, pp. 175, 295–393.

Jacobs, R.M., The technique of design review, in *Proc. of the 1979 Product Liability Prevention Conference*, American Society for Quality Control, Milwaukee, WI, 1979, p. 80.

Jacobs, R.M., Evolution of quality/reliability due to litigation, in *Proc. of the 1983 Reliability and Maintainability Symposium*, Conference Committee, Orlando, FL, 1983, p. 122.

Jacobs, R.M. and Mihalasky, J., Practices and systems for product liability prevention, in *Proc. of the 1976 American Society for Quality Control Technical Conference*, American Society for Quality Control, Milwaukee, WI, 1976, p. 105.

U.S. Consumer Product Safety Commission, *Handbook and Standard for Manufacturing Safer Consumer Products*, U.S. Government Printing Office, Washington, D.C., 1975, pp. 10–16.

7 Avoiding Product Liability: Manufacturing and Use

Introduction

The product has been designed and the designs tested. Suppliers have been evaluated and selected. This chapter offers specific tools and techniques to be employed in the remaining stages of the product life-cycle. Consideration of these methods is encouraged to foster the Financially Focused Quality mindset and to ensure the quality and safety of products.

Inspection and Test of Vendor Components

Critical decisions to be made by the prime manufacturer include how much inspection and testing of a vendor component should be done and the method(s) to be employed for the inspection or testing. The only way to be 100% sure that every item meets required specifications is to test every single item and product on the vendor's assembly line (Eginton, 1976). In the cases of some highly critical products, this is feasible and is actually accomplished. However, the extent of testing is a decision that must be examined carefully from the quality, safety, and cost standpoints.

The four primary methods that may be used for the inspection and testing of vendor components are first article inspection, source inspection, vendor inspection, and receiving inspection. These methods are discussed below.

First Article Inspection

The very first item produced by the supplier on the production line is usually given an intensive evaluation, often called the first article inspection, which provides a basis for determining the supplier's ability to duplicate the product under normal manufacturing conditions. This inspection may include environmental and destructive type tests.

Source Inspection

Inspection performed at the supplier's plant by a representative of the prime manufacturer is known as source inspection and is used in the following situations:

1. Adequate inspection at the purchaser's facility by supplier personnel is prevented due to the level of the assembly.
2. Inspection or test of only completed products does not adequately determine the quality of the manufacturing processes.
3. Required environment or test equipment is not available at the purchaser's plant.
4. The items are to be shipped directly from the supplier to the purchaser's customer or to an outside facility.
5. It is more economical to inspect at the supplier's plant.

In addition to completing the usual inspection tasks, the following attributes of competent source inspectors have been suggested (Wilson, 1980):

1. *Diplomacy:* When in the supplier's facilities, the inspector not only represents the company but he also is usually the only person apart from the buyer that the supplier sees. His actions and statements reflect directly back to the company.
2. *Salesmanship:* Particularly where rejects are involved, the inspector must be able to "sell" the supplier on the fact that the

rejection is valid and logical, while at the same time convincing the supplier to accept the rejection. This is especially true of a borderline rejection.

3. *Ability:* The inspector must be able to work virtually without supervision. In most instances, the source inspector is many miles from his company, and in some cases could be in another state. He cannot always consult with his supervisor or other persons within the company to help him make a decision or resolve a supplier's objection to his decision.

4. *Appearance:* The inspector should be neat, clean, and well groomed. He not only represents the company; to the supplier, he *is* the company. Psychologically, more authority is perceived when an inspector wears suitable business attire, as opposed to jeans and a t-shirt.

Vendor Inspection

If, for economic or other significant reasons, a manufacturer feels it is impossible or impractical to perform source inspection or to perform receiving inspection at his own plant, vendor inspection may be the only assurance of component quality. When this is the case, the manufacturer should require that the vendor supply evidence of such inspection and testing. In this situation, it is particularly important that the manufacturer perform periodic audits and follow-up audits of vendor operations.

Receiving Inspection

Receiving inspection is used to ensure that purchased items are in conformance with the design documentation. Even when items are source inspected, the purchaser should inspect them upon receipt. This is necessary to detect damage that may have resulted during transport from the supplier.

As a component of product liability prevention, the Receiving department should retain records of when items were received from suppliers, how they arrived, the purchase orders, and the specifications for the materials or parts received. In addition, the records of inspection should show the quantity received, the number of units inspected, the date of issue of the specifications and drawings against which the parts or materials were compared, the number of units accepted, and the disposition of those rejected. Specific attention is required for those items returned to the supplier, repaired, or reworked and those that were accepted "as is" or were handled by some other disposition (Jacobs and Mihalasky, 1976).

Inspection Records During the Receiving Cycle

The results of inspections made during the receiving cycle should be recorded in a file in such a manner that the history of a supplier's performance on a specific part can be readily determined (Cound and McDermott, 1971). The data contained in this file can be quite valuable. For example, supplier performance data are useful when selecting new suppliers and also aid when selecting the appropriate level of inspection (i.e., reduced, normal, or tightened). Unfavorable trends and problems that may be contributing to current detrimental situations can be identified from these records. Specific part quality levels are useful in tracing past problems and developing corrective actions.

Source/Receiving Inspection: Corrective Action

Many problems identified during source inspection or receiving inspection require immediate notification of the supplier and corrective action. Periodic reviews of the inspection records may reflect the existence of a more subtle quality problem or an unfavorable trend in supplier performance. When corrective action is required, the supplier should be notified in writing with sufficient detailed information to identify clearly the nature and magnitude of the problem. The supplier should also be requested to respond with a plan of corrective action by a specific date.

Inspection of Storage Facilities

Items passing source or receiving inspection are usually stored in storage facilities until required by the manufacturing organization. These facilities must be inspected periodically to ensure that the items are not degrading during the storage period. The following factors should be checked during such inspections (Cound and McDermott, 1971):

1. Proper storage and identification to prevent commingling and loss of items
2. Proper packaging and shelving to prevent damage from atmospheric exposure, excessive weight, or related factors
3. Proper rotation of stock per specific provisions to prevent excessive aging and deterioration of parts
4. Proper control of items having specific shelf-life requirements (for example, certain rubber goods, plastics, chemicals, etc.)

Inspection and Testing in Manufacturing

The primary reason for performing inspection and testing is to determine whether or not products conform to the specifications of the design documentation. This purpose is often called "product acceptance" or "acceptance inspection" (Juran, 1988).

The Obligation To Inspect and Test

Borel v. Fibreboard Paper Products Corp., 493 F. 2d 1076 (5 Cir. 1973) established that a manufacturer has the legal duty to inspect and test his product (Greco, 1980). When attempting to minimize liability exposure, companies need to be aware that the obligation to inspect must vary with the nature of things to be inspected. There are many situations when obligation exists. For example, a defendant can be charged with failure to inspect when defects could have been discovered by reasonable inspection.

A defendant can also be held to have failed its duty of care to the plaintiff in negligently assembling, testing, and inspecting the product. In addition, the manufacturer or the seller owes the duty to inspect the product knowing that the purchaser relies upon it being done. Statutory command requires that products under the jurisdiction of regulatory standards be tested and inspected. Products might also be inspected in conformance with specific contractual requirements of safety agencies.

Inspection and Test Areas

In order to allow for safe and accurate inspection and testing, these functions must be performed in specially designed areas. Such areas should be planned by inspection and test supervisors, quality control engineers, and manufacturing planning engineers, and their efforts must be thoroughly coordinated.

The first factor to consider when planning such areas is whether it is best for the inspector to come to the work or vice versa. This decision is directly influenced by the work flow, the inspection or test criteria, and cost. Juran (1988) points out that in receiving inspection, the problems of transport and unloading of the incoming product usually dictate the need for a common area for receiving all incoming shipments. It becomes logical to locate the receiving inspection in this same common area, often with specially designed systems of transport to facilitate sampling and disposition. For the in-process inspection functions, Juran also recommends that the inspector go to the work. The wide dispersion of the production areas and the sheer mass of material involved would make it very difficult to take the work to the inspector.

The final inspection function most likely is performed in a specific area, preferably near the shipping dock. Usually this is expedient because of the presence of permanently installed test facilities and power sources. Another reason for performing final inspection near the final inspection or testing area is that very little product movement is required for packaging and shipping.

Measurement

The design documentation specifies attributes that the product is to have. Often it is necessary to use various measuring tools to determine how closely the product or components thereof conform to these requirements. The four basic characteristics of measuring equipment are

1. *Sensitivity,* which pertains to the degree to which measurement variation can be observed
2. *Precision,* which reflects the degree to which an instrument will reproduce a given measurement under constant conditions
3. *Accuracy,* which is the degree to which an instrument measures the true value as represented by an accepted standard, such as those provided by the National Bureau of Standards
4. *Repeatability,* which refers to the ability to obtain the same measurement of a given quantity if measured more than once

Sensitivity and the precision of instruments have an important bearing on the selection of specific instruments for particular measurement requirements. Accuracy is generally maintained through periodic calibration of the instrument to standards of known accuracy (Cound and McDermott, 1971). Described below are characteristics and basic factors that should be considered when selecting measuring instruments for particular uses:

1. *Type of instrument:* Indicating instruments, such as micrometers and dial indicators, provide actual readings which show where the process is set and how much variation is occurring. Fixed instruments, such as plug gaps, ring gauges, and adjustable snap gauges, show whether a part is undersized, within limits, or oversized.
2. *Degree of sensitivity:* Generally, the instrument should be sensitive enough to permit dividing the total allowable range of the characteristic into tenths.
3. *Degree of precision:* The degree of instrument precision may be determined by making a number of measurements under the same conditions on the same unit of the product; these values are then used to calculate the standard deviation of the instrument, which can be stated in terms of the instrument tolerance.

Many manufacturing concerns establish their own standards laboratory where they maintain the company's reference standards. The reference standards are the highest accuracy devices in the company, and their calibration requires special techniques and extra time. Reference standards must be maintained under highly controlled environments, which, in turn, dictate special facility requirements.

Training of Inspectors

From the first day inspectors assume their new responsibilities, a concerted effort should be made to instill in them the Financially Focused mindset. The orientation of new inspectors should include the following:

1. Introduction to the organization
2. Tour of facilities

3. Detailed description of the job
4. Training on the equipment that will be used
5. Instruction on tracings, Quality Assurance procedures, and what is expected of an inspector
6. Encouragement regarding the necessity of having a curious and investigating mind and maintaining free communication with supervisors and business communication with production associates

The Quality Assurance inspector should, when possible, become familiar with the end product as a technically qualified individual and directly question probabilities of liability (for example, sharp edges, marginal workmanship, inconsistent quality, and quantity of inspected product). The inspector should immediately be trained in use of the company's Process Improvement Recommendation (PIR) form and be offered an incentive for its use. Three areas that should be emphasized by all manufacturing concerns include:

1. Ongoing inspector training (on-the-job or formal)
2. Inspector attendance at seminars, technical courses, and conferences
3. Membership in professional societies such as the American Society for Quality (ASQ)

Inspection Instructions

Many companies use detailed inspection instructions for specific products. These instructions usually include information of the following nature (Cound and McDermott, 1971):

1. A listing of the tools to be used in the process
2. Summary of specific characteristics to check
3. Detailed descriptions of special techniques or methods which are required
4. Sampling tables
5. Reference guides
6. Visual aids
7. Directions regarding frequency and type of samples to be forwarded to laboratories for various tests

Inspection Procedure Manuals

Juran recommends that inspection instructions be included in a procedure manual, whose contents consist of the following items (Juran, 1988):

1. Statement of legitimacy and purpose, approved by the responsible manager
2. Table of contents of the manual
3. Organization section, including inspection organization charts, job descriptions, and statements of responsibilities
4. General concept of inspection as used in the company.
5. Plan for seriousness classification of defects (actual classifications of defects are in the supplemental product manuals)
6. Standard plans for sampling inspection: bulk sampling, tables of random sampling, continuous sampling, narrow-limit sampling
7. Standard plans for use of control charts
8. Vendor inspection procedures
9. In-process inspection procedures
10. Finished goods inspection procedures
11. Measurement control procedures, including the schedule of checking intervals for general-use equipment
12. Copies of all the inspection forms used and instructions for data recording and documentation
13. Product identification procedure
14. Procedure for issuance and control of inspection stamps, which are used by inspectors to identify the inspection status of items
15. Feedback of data to production and procedures for corrective action
16. Procedures for dealing with nonconforming material
17. Index and glossary

Such instructions must be controlled, and changes in product specifications must be reflected in corresponding changes to the inspection instruction. Lighting conditions under which inspections and tests are to be performed may also be designated. Studies have shown that lighting conditions dramatically affect the performance of inspectors (Peterson, 1980).

Industrial Quality Assurance

Production involves utilization of various types of processing equipment (tools, machines, measuring equipment) and the actual physical labor of machine operators. The goal of this process is the production of end items that meet or exceed the safety and quality characteristics of the design documentation.

Assessing the Production Process

In a dynamic, complex industrial operation, it is seldom possible to evaluate the production process fully prior to initiation of production. In most cases, the nature and type of process evaluation develop with time. The two basic types of process assessment techniques that are utilized are described below.

Process Capability Studies

Such studies are used to determine the inherent capability of specific elements of the production processing. For example, studies may be made to evaluate a chemical processing operation, a console of interconnected items of electronic test equipment, a measuring device, or the capability of individuals to perform specific operations.

Mandatory Process Evaluation

It is virtually impossible to evaluate every variable in the production process. Thus, most companies establish mandatory process evaluation points for those variables that are critical to achieving safety and quality objectives. Process evaluation points may include the methods described below:

1. *Operator certification:* When critical skills (soldering, welding, etc.) are required, operator certification training programs are used. After taking specific classes, operators are required to pass certification exams and, thereafter, periodic recertification exams.
2. *Tool proofing:* Tools and fixtures used in the production process must be evaluated and approved prior to their use in the production process.

3. *Critical processes:* Critical processing operations, such as heat-treating, plating, etc., require evaluation and approval prior to use.
4. *Test operations:* Complex testing operations frequently require review and approval to ensure accurate testing of the product regarding particular functional test specifications.

The Manufacturing Process

The basic function of the manufacturing organization is to convert raw materials into finished products. The functions for preparing the product for delivery are also within the manufacturing realm of responsibility, unless a specific logistics organization has been established. The manufacturing process may include the following:

1. Processing
2. Machining
3. Fabricating
4. Assembly
5. Packaging
6. Packing
7. Shipping

The required product and process characteristics are established in the design documentation. The Quality Assurance function is instrumental in ensuring that manufacturing operations are performed properly and that the intended characteristics exist in the finished product.

Process Control

An overall objective of the Quality Assurance function is the prevention of defects. This goal is best achieved through effective process control. If the process is statistically controlled at the level specified for a certain characteristic, there should be no defects produced for that characteristic. Variations in the production process result in variations from product to product. Statistical quality control recognizes that no two products will be exactly identical. It also recognizes that variability follows a characteristic pattern for a given process and that the variation should fall within certain limits.

In-Process Inspection and Testing

Inspection and testing performed in-house during the manufacturing cycle is known as in-process inspection and testing. These functions are performed to ensure that the finished products will perform their intended functions safely and to improve yields at the end of the process. Consumer products must be inspected and tested prior to distribution in order to verify their conformance to established requirements.

If a product is made up of components or subassemblies that are not accessible for inspection and testing, these functions must be performed, as applicable, before such items are assembled into parent units and become inaccessible. It is the responsibility of the manufacturer to provide guidance for inspection and testing to the degree that operators are fully informed as to how to conduct inspections and tests that are meaningful, objective, and uniform and how to record the results. Even with the best planning and prevention activity, errors are still going to occur during the production process. Inspection and testing are used to detect errors and provide for the feedback of information for the correction of the basic cause of these errors.

Once the decision has been made that inspection and testing are required, four basic questions must be answered:

1. What is going to be inspected and tested — the operation producing the product or the product itself?
2. When will the inspection and testing be performed (preferably prior to very expensive processing operations)?
3. Where will the inspection and testing take place (at which product locations, in what controlled environments, etc.)?
4. How will the inspection and testing be performed (instructions)?

Safety and quality verification is usually accomplished by the use of either destructive testing or nondestructive testing of raw materials, components, or finished products. Destructive tests permanently alter the shape and/or characteristics of the item. Such tests may be performed during development testing of the product to determine the limits of product capability. These tests may be performed on a percentage of items (sample inspection) to ensure that product characteristics are acceptable. Nondestructive tests determine whether the manufactured product meets manufacturing specifications without altering the product. Nondestructive tests may also be used to determine if, due to degradation of safety critical characteristics, products in the field require repair or replacement.

Nondestructive tests may play an important role in failure analysis and the litigation of product liability lawsuits. These tests can be used to determine if a defect exists in a weld or whether a casting contains excessive voids. The most widespread method of nondestructive testing is visual inspection. The inspector uses either his eyes alone or special instruments to determine if an item in question has been manufactured to specifications.

Other methods of nondestructive testing involve industrial radiography, liquid penetrant, ultrasound, magnetic particles, and eddy currents. Greater detail on these and other methods of inspection can be found in *Quality Assurance: Management and Technology* (Hayes, 1982).

Tolerances

No two manufactured articles are exactly alike. Variability is one of the fundamental factors influencing approaches to Quality Assurance. Tolerances are established for quality characteristics to allow for a limited degree of imperfection in the manufactured product. Tolerances should not be set so loose that the product tends to function improperly and/or be unsafe. In contrast, needlessly tight tolerances could lead to unjustifiably high manufacturing costs, as manufacturing costs usually rise as tolerances tighten. Financially Focused Quality often calls for the Process Improvement Follow-Up Plan to consider a loosening of tolerances after tests have been conducted and such actions are justified.

Nonconforming Material

In most manufacturing operations, for one reason or another, material will occasionally fail to conform to established requirements, and measured deviations will exceed the established tolerances. Such nonconforming material not only results in products of unsatisfactory quality but also is a potential hazard to safety because it can be assembled easily and inadvertently into end products. Therefore, it is essential that nonconforming material be clearly labeled and segregated. Once nonconforming material has been so identified, it must be given a disposition.

A disposition gives directions for how nonconforming material should be handled in the future (for example, rework, scrap).

Quality Assurance Through Statistics and Sampling

Statistics refers to the science of collecting, organizing, and interpreting numerical data. In the quality control sense, the interpretation of this data leads to decisions regarding the safety and quality of products. Economy of inspection is achieved by not inspecting an entire lot to determine lot acceptance or rejection. Thus, manufacturers for inspection, testing, calibration, and process control generally use statistical techniques when possible. Cases where statistical methods may not be appropriate include products with critical characteristics or when pertinent standards require the inspection and testing of each unit or product.

Basically, there are two types of statistical data: those that are counted and those that are, in the usual sense, measured. Counted data are called "attributes" and are represented by whole numbers — for example, the number of items passing a test. Measured data can be considered as "variable" and having an infinite number of values. Such data can be presented on a scale or continuum for measurement of time, distance, pressure, temperature, and the like (Hayes, 1982).

With measured data, the repeatability of the measurement tends to increase the variance of the data, while the accuracy of the measurement tends to dislocate the central tendency. Statistical analysis begins with the collection, tabulation, and classification of data. Worksheets are generally employed to ensure a systematic method of handling data. An array is data arranged in order of magnitude (ascending or descending). Data should be arranged in an array to have relevant meaning.

Definition of Terms

Abbott (1974) presented the following four definitions of terms to promote a closer understanding between the legal and technical aspects of sampling:

1. *Standard* is a description and specification of product characteristics which affect safety. This includes (a) a clear definition

of each safety characteristic and the hazard that it is intended to prevent; (b) a quantified measuring system for each of the definitions, such as scale with defined units of measure; (c) specification limits in terms of the measuring system, such as points on the scale indicating minimum and/or maximum values defining permissible zones of safety; and (d) a system of methods and procedures to evaluate conformance to the specification limits, such as a standard test procedure.

2. *Compliance* is conformance of a product to all applicable specifications in a safety standard.

3. *Certification* is a statement by the manufacturer that the units of production covered by the certificate are in compliance with the applicable stated standards.

4. *Enforcement* is the totality of activities by the Consumer Product Safety Commission in assuring compliance of products to safety standards.

Sampling Inspection

Statistical sampling is one of the most useful tools that the quality control and safety engineer has at his disposal. While statistical methods are used for many different purposes (for example, charting process trends), their most common application is to lot-by-lot sampling by attributes. Random samples are drawn from a lot or batch and inspected for specific characteristics. Each unit is then classified as acceptable or unacceptable for the characteristic. In addition to lot-by-lot inspection, continuous sampling inspection is often employed. In continuous sampling inspection, current results determine the type of inspection for the products to be inspected.

Data are also used to ascertain trends in various parameters. For instance, data can indicate that a dimension is becoming too small due to tool wear, and this problem can be fed back to the supplier before quality becomes unacceptable and production is stopped.

Random Numbers

All sampling plans, other than 100%, are based upon the random selection of the sample. This means that all items in the lot have an equal chance of being selected in the sample. A table of random numbers

80	38	81	69	45	87	17	80	00	50	91	77	82	23
84	48	01	07	15	80	44	74	25	29	63	07	56	11
11	77	17	61	15	47	44	98	48	27	33	51	52	80
52	66	94	55	72	85	73	49	14	46	26	25	22	96
67	89	75	43	87	54	14	63	05	52	28	25	62	72
90	56	86	07	50	15	14	33	52	74	41	89	28	66
48	14	46	42	75	67	88	45	13	87	46	75	89	45
96	29	77	88	22	54	38	09	12	00	87	69	99	33
21	45	98	91	49	91	45	03	67	19	95	12	30	18
53	68	47	92	76	86	46	24	95	52	63	18	29	01
16	28	35	54	94	75	08	93	95	24	25	66	59	94
99	23	37	08	48	32	47	05	17	58	64	23	11	45
79	28	31	24	96	47	10	64	26	37	65	41	10	78
02	29	53	44	01	41	51	78	39	84	56	85	97	91
82	16	15	01	84	87	69	38	76	36	16	45	49	65
99	38	94	99	00	03	15	06	59	33	70	32	79	24
46	75	75	30	32	70	68	35	98	51	17	62	13	44
53	29	02	10	47	96	24	63	55	18	98	48	41	13
31	28	79	47	32	48	56	15	90	27	66	73	70	62
22	33	12	46	88	40	34	72	29	33	06	41	38	38
33	97	09	39	15	17	38	07	41	76	80	73	70	79
33	35	13	54	62	83	10	36	48	24	56	94	38	69
29	95	72	28	64	05	18	07	49	41	38	87	31	24
27	89	34	20	24	63	47	17	18	15	65	86	73	80

Figure 7.1. Random Number Table

(Figure 7.1) is frequently employed towards the goal of ensuring randomness.

All sampling inherently involves risk. Most of the standard sampling plans identify risks by operating characteristic curves. These curves are used to display the probability of acceptance or rejection at various levels of number of defects found (for example, 1% or 5% defective). Receiving inspection operations utilize sampling inspection extensively because it allows the processing of a large number of items with a high level of assurance at a minimum cost.

Sampling plans are further classified depending on whether the quality characteristics are measured and expressed in numbers (i.e., variable inspection) or whether articles are classified only as defective or not defective (i.e., attributes inspection). An alternative to sampling inspection is to screen every item (100% inspection), but this approach is rarely cost effective.

During subsequent manufacturing operations, the purchased items will receive additional checks, thereby reducing the risk of a defective item being included in the finished article. The costs and losses associated with the detection of this defective item later in the production process must be balanced with other considerations in determining the acceptable risk level and associated acceptable quality levels. Financially Focused Quality offers tools that enable the achievement of an appropriate balance.

Sampling Plans

A sampling plan specifies the following:

1. Sampling plan to use (e.g., single, double, or multiple sampling)
2. Size of samples based on the lot size
3. Acceptance criteria (e.g., the number of permissible defects in the sample)

Each sampling plan has stated risks. The quality control engineer must be thoroughly familiar with the operating characteristics of the plan being used so that he will know the risk of delivering defective products.

Single sampling is the most easily administered and by far the least complicated as far as procurement of samples is concerned. A single sampling plan is one by which the inspection results of one sample yield judgment criteria for either accepting or rejecting a given lot. Double sampling plans are similar to single sampling plans, except in this case judgment is based on the inspection of one or two samples. The second sample is inspected if the number of defects found in the first sample is less than the rejection minimum but greater than the acceptance minimum. Depending upon the accept/reject criteria, many samples may be drawn before a decision is made regarding accepting or rejecting a lot.

The maximum number of defects per hundred units (percent defective) that can be considered satisfactory as a process average is known as the acceptable quality level (AQL). Most quality/reliability specialists use the term AQL, believing that if the number of allowable defects is small enough, an adequate job is being performed. Jacobs (1983) points out that the various U.S. courts do not agree that a small AQL is permissible. They are requiring that defendants prove that the single device that failed, causing an injury, left the manufacturer's control in a nondefective condition. In other words, courts are requiring justification for failure of the device in question, even if over 1000 were produced without failure resulting. The probability is low, but for the injured individual, the courts have ruled that each unit must be evaluated by itself.

It is important to stress that under no circumstances should a single failure endanger human life. Product designs should be fail-safe. The product must be manufactured to conform to such fail-safe designs, and the inspection process must ensure that product safety does indeed exist.

When submitted quality is considerably worse than the AQL, double and multiple sampling require less average inspection because of rejection on the first sample. When quality is intermediate, double and multiple sampling may require more inspection than single sampling.

Volumes and volumes of mathematical tables, formulas, and theories have been developed to support sampling and statistical quality control functions for virtually every type of manufacturing concern. More detailed introductions to statistical methods of quality control can be found in many books at public libraries.

Degree of Inspection

The degree of inspection utilized by the manufacturer is highly dependent on historical quality level as compared to AQL. For example, tightened inspection provides protection against acceptance of lots that have quality slightly below the AQL. Similarly, reduced inspection permits a reduction in the inspection force when historical quality is consistently higher than the AQL. However, reduced inspection should be avoided, as protection against the occasional bad lot will be greatly reduced. Financially Focused Quality encourages an analysis (for example, corrective action analysis of cost) when a company is investigating the potential for changing inspection levels.

Sampling Plans vs. 100% Inspection

Sampling plans are often preferred over 100% inspection in certain cases for the following three reasons:

1. Many comprehensive tests are destructive. If 100% of products were destroyed in testing, there would be none left to sell. On

the other hand, inspection or nondestructive testing often cannot completely measure the safety parameter.
2.   Because many tests or inspections are complex or time consuming, the inspectors would tire, leading to increased risk of human error.
3.   One hundred percent test or inspection is not needed to assess the variability of the safety parameter. For mass-produced products, 100% inspection would increase costs. Increased costs would be reflected in increase retail prices that the consumer would be unwilling to pay (Abbott, 1974).

Conversely, 100% test or inspection can be effectively used when tests or inspections are nondestructive and the procedures are simple and short, thus minimizing human error; or, automatic inspection or testing can be incorporated into the manufacturing process, where benefits justify costs to eliminate most human error.

Arguments Against Sampling Plans

There is much justification to support the use of sampling plans to ensure consistent quality; however, three reasons why a manufacturer may decide not to use such plans are presented below:

1.   Sampling plans, in practice, will always allow for intolerably large deviations from whatever performance standards may have been set.
2.   Sampling plans, particularly those based on attribute sampling, cannot forecast the magnitude of danger inherent in nonconforming items within an acceptable lot.
3.   Use of samplings will give the public a false impression of greater quality control than will actually exist.

Inspector Errors

Even in the case of 100% inspection, and even if the tools being utilized are perfectly calibrated, there will be an occasional defect or two that is not spotted by the inspector. The human element in the inspection process is an important factor, as it contributes to inspection errors.

There are basically three categories of inspector errors: (1) technique errors are caused by the lack of capacity, skill, or know-how; (2) inadvertent errors are those deemed unintentional and should not occur; (3) willful errors might result from the actions of a disgruntled employee. Management must review carefully the quality, safety, and cost impacts when deciding between sampling plans and 100% inspection.

Consumerism

The product has completed the manufacturing process. The company must now get the product to the customer and guarantee his satisfaction.

Shipping Inspection

With the exception of field sampling at the point of sale or at the customer's site, shipping inspection is the very last opportunity the manufacturer has to examine his product before it reaches the ultimate consumer. During this phase, the inspector should verify the following (The Travelers Insurance Companies, 1973):

1. The product functions properly (to the extent practical)
2. The product is properly packaged (see below)
3. The product or packaging bears the correct labeling/nameplating
4. Associated warnings are included.
5. Instructions for use are provided.
6. Warranty and/or guarantee is included.
7. The product has markings or seals required by regulatory agencies.

Packaging for Product Safety

To ensure that specified quality and safety levels are maintained en route from the manufacturer to the customer, the product must be properly packaged. When shipping and packaging containers are being designed, special product protection requirements must be considered. Such requirements may include protection against dust, microorganisms, extreme temperatures, humidity, sunlight, vibration, air, shock, abrasion, and water damage. Special handling and packaging can provide the necessary protection. Such protection may involve hermetically sealed packages, refrigeration, desiccation, grease or plastic coatings, special wrappings, and even floating packages. Packaging, like the product itself, must undergo tests to ensure its protective ability (Cound and McDermott, 1971).

Product Safety in the Field

Obviously, the majority of litigations resulting from defective products occur when the product is in the hands of the consumer. Many factors influence the safe use of the product by the customer, the primary factor of which is the service publication.

Service Publications

Service publications are instructions for operation or service of the product by the consumer. Product safety is largely influenced by the quality of the instructions. Their adequacy and completeness are essential. The following techniques may be used to ensure high quality in these documents:

1. Document is validated by a demonstration of its use. With this method, the instructions are actually followed on a step-by-step basis to detect errors or omissions.
2. The document is reviewed by an independent group or an individual not intimately familiar with the product.

Field Results

Because safety in the field is a primary goal of the quality assurance program, a method of measuring the degree of experienced customer safety should be established. Customer injuries and complaints provide a meaningful measure of danger, which is an inverse measure of safety. Records of such field data should be carefully examined and followed up, whether or not a lawsuit is pending.

Sometimes it is too expensive to process all reports from the field. In such situations, a sample of the complaints may be examined; however, all injuries and most safety-related complaints require careful examination. The following steps should be followed per the framework of Financially Focused Quality:

1. The Failure Notice must identify the specific injury/complaint.
2. Potential failure analysts determine possible causes (product defect, improper use, etc.).
3. Failure analysts with the Financially Focused Quality mindset (and with a representative of Finance, if appropriate) should perform analyses towards accomplishing the following:
   a. Determine specific cause(s) of injury/complaint.
   b. Recommend potential process improvements (for example, design or production process change, product instruction manual rewrite, etc.).
   c. Determine the cost impacts associated with each potential process improvement.
   d. Select the most cost-effective alternatives.
   e. Develop a Process Improvement Follow-Up Plan (PIFP).
4. Follow-up must be performed in accordance with the PIFP.
5. A process improvement closure notice is issued when the PIFP has been satisfied.

The very best way to perform in-depth analysis on field failures is for quality engineers to have direct access to the failed product. This is why FFQ recommends that both the Failure Notice and the failed component be forwarded to the process improvement coordinator, who, when possible, should request that the product involved in the injury or complaint be sent to him at the factory for hands-on investigation. The eventual process improvements related to field failures may include improved design, improved test and/or inspection methods, better packaging, or better instruction books.

Product Recall

Another function in which the Quality Assurance organization should be involved is product recall. Representatives from the Finance, Engineering, Manufacturing, Quality Assurance, Purchasing, Sales, Marketing, Advertising, Legal, and Insurance departments should serve on a central committee prepared to take specific action in the event that flawed products are being distributed to consumers.

It should be obvious that no company should begin marketing a product anticipating its recall. However, to be adequately prepared, a company should assign members on the recall committee the responsibility of establishing formal procedures for handling such situations. The routine should be completely spelled out before the need arises.

Management Resources for Assuring Quality

Management has many sources available to ensure that the company is manufacturing safe, reliable, and quality products.

Accredited Laboratories

Sound product liability loss prevention techniques minimize the chances of a defective product being manufactured. One such technique is the use of product designs that rely upon well-accepted standards and are backed by quality assurance programs, which incorporate tests made by competent laboratories. In this case, a "competent laboratory" has been defined as being a laboratory for which the competency is attested to by well-recognized governmental or private-sector accrediting organizations (Forman, 1978).

As shown in the Underwriters Laboratories, Inc. (UL) case discussed in Chapter 6 in the Responsibility for Product Safety section, employment of an accredited outside test agency does not guarantee that a product will be free from litigation. However, the advantages of such employment do merit discussion.

Underwriters Laboratory has established itself as a leader in the field of testing products for safety, having performed this function for 90 years. Harvey Berman, Associate Managing Engineer for UL, explained that UL investigation of a product consists of an evaluation by means of tests and examinations to determine compliance with the requirements contained in the UL Standard for Safety. Berman further explained that such requirements generally emphasize the ability of a product to perform safely under its intended use (Berman, 1984).

Howard I. Forman, Deputy Assistant Secretary for Product Standards for the U.S. Department of Commerce, believes that a key factor in the manufacture of good quality products can be the competency of the laboratories that perform tests on components and end products (Forman, 1979). The same devices used to assure the competency of a supplier (see Supplier Selection section in Chapter 6) may be used in regard to outside test laboratories.

Similarly, a manufacturer may choose such a laboratory for many of the same reasons as those for choosing a supplier. Another reason for utilization of an accredited laboratory may be related to promoting a particular product. Customers of the manufacturer may be made aware of the tests, and, if test results are favorable, customer confidence may increase, leading to greater interest in buying the product.

Also, the use of an accredited laboratory provides added support for good manufacturing practices in defense of product liability suits. In jurisdictions where evidence of such practices is admissible and given significant weight if fairly established, each practice will have to stand on its own and be judged accordingly.

Audit Control

Even the best quality assurance program is vulnerable because of the human factor involved. It should be obvious that a system of checks and balances is necessary. After all, why should litigation regarding a defective product be brought about if the defect could have been detected prior to leaving the factory?

The Travelers Insurance Company (1973) reported that an effective check-and-balance tool is audit control of all animate and inanimate functions that could contribute to bad products. The type of product, amount of production, and procurement arrangement will dictate the frequency and depth of audit control and the size of the team. The important point is that all audits should be conducted as an addition to routine reviews, inspections, and tests. To be an effective management tool, they should be formally structured with written results. Effective audit control includes software and hardware audits in the areas of quality assurance programs and procedures, company policies, and configuration of primary and secondary control documents.

Vendor Audit

After the vendor has been selected and is supplying components, audits should be conducted to evaluate performance according to the previously agreed upon control systems. The audits should evaluate both the system and the product, and they should be conducted at random intervals. Such audits may also be conducted at regular intervals, but there is a lower likelihood of significant audit findings if the vendor knows in advance exactly when the audit will take place.

Each audit should be performed with adequate preparation. A well-prepared checklist may provide the basis for a thorough supplier evaluation. The following checklist is offered as a sample of the elements that a vendor audit should include:

1. Dimensional conformance
2. Records of inspection and test with lot traceability
3. Written decisions regarding nonconforming material
4. Corrective action system (for example, Financially Focused Quality)
5. Evidence of the occurrence of internal audits
6. Notification of nonconformances
7. Operator job knowledge
8. Management knowledge of requirements
9. Attitudes of management, staff, and line employees
10. Calibration records
11. Compliance with product safety guidelines from applicable industry or government standards

After each audit, the results should be discussed with the supplier. During this discussion, corrective action dates should be agreed upon and an audit follow-up date should be established. The purpose of the follow-up is, of course, to ensure that corrective actions have been taken and that the supplier is performing within the agreed-upon control standards.

Product Safety Audit

A manufacturer must establish, document, and implement practices and procedures to ensure that the product is safe and that it is manufactured safely. It is then critical that these practices and procedures are followed. To ensure compliance, prudent management will initiate an ongoing program of safety audits.

Such a program should examine practices and procedures to ensure they are formally documented with job instructions. Job instructions should adequately address the specific requirements of each individual product. Steps are required to ensure that the actual practice conforms to the documented job instructions.

The program should effectively improve product safety and reduce product liability exposure. The successful product safety audit begins with the preparation phase. Careful preparation is followed by the execution, where the audit is performed. During the third phase (reporting), the auditor presents the audit findings to those that were audited as well as those concerned about the outcome (for example, management). The last phase (follow-up) involves closure to ensure that reported deficiencies have been resolved.

During the preparation phase, the auditor needs to understand the purpose of the audit. With the purpose in mind, the auditor should plan the type of audit to be performed. There are four primary types of product safety audits, each focusing on a different area. These audits encompass the following areas:

1. Systems
2. Processes
3. Products
4. Procedures and methods

Quality Assurance Evaluation Checklist

The following checklist of questions is offered to manufacturing management as a starting point for areas to consider when evaluating the quality assurance systems in place (Cappels and Bass, 1986):

1. Are quality assurance actions instituted throughout manufacturing to prevent and detect product deficiencies and safety hazards?
2. Are inspections and tests conducted in accordance with written plans?
3. Do these plans contain explicit inspection and test instructions?
4. Are new quality assurance employees provided with formal training and certification prior to hands-on involvement in the process?
5. Are inspections and tests conducted in the manufacturing flow before potential safety hazards become inaccessible for detection?
6. Are sampling plans adequately defined and described?
7. Are the risks acceptable for the applications?

8. Is 100% inspection or testing required for characteristics that are potential safety hazards?
9. Are the disciplines required for scientific sampling enforced (for example, random selection, strict compliance with decision rules, records of results)?
10. Are formal quality assurance procedures and job instructions routinely reviewed or updated as required?
11. Are periodic internal product safety audits conducted to ensure compliance with formal procedures and job instructions?
12. Are nonconforming materials and products distinctly identified?
13. Are nonconforming materials and products segregated from conforming materials and products?
14. Are the dispositions of nonconforming materials and products accomplished as prescribed?
15. Is the policy for controlling measurement and calibration instruments clearly defined?
16. Is there a central location where procedures, job instructions, definitions, explanations (sampling plans, etc.) are available?
17. Is there an adequate minimum ratio of instrument and reference standard accuracy to the established tolerances?
18. Do detail procedures adequately identify responsibilities, calibration and inspection intervals, records, due dates, and frequencies for tools and equipment to again be calibrated?
19. Are reference standards used which are traceable to a valid source?
20. Is the manufacturer aware of standards currently being applied in the industry?
21. Are protective provisions adequate?
22. Are arrangements made to ensure that suppliers comply with calibration requirements?

Summary

Prudent management has many tools at its disposal to avoid costly litigation. A true financial focus ensures that intelligent decisions are made throughout the product life-cycle and that costly litigation is extremely rare.

Self-Study/Discussion Questions

1. In what ways does consideration of potential product liability issues add to the costs of a company? How can such a consideration decrease costs?
2. Share your experiences with product recalls. How much do you think it costs a company for product recalls? How could such failures have been prevented?
3. What is your opinion of the instructions that accompany products — for example, assembling a bicycle, operating a VCR, setting the time on a digital watch? Do you think adequate guidance is provided?

References

Abbott, R.A., Sampling — the right way — for liability prevention, in *Proc. of the 1974 Product Liability Prevention Conference*, pp. 173–174.

Berman, H.S., Product safety and quality assurance at UL, in *Proc. of the 1984 American Society for Quality Control Quality Congress*, American Society for Quality Control, Milwaulkee, WI, 1984, p. 123.

Cappels, T.M. and Bass, L., *Products Liability — Design and Manufacturing Defects*, West Group, St. Paul, MN, 1986, p. 294.

Cound, D.M. and McDermott, T.C., *The Handbook of Industrial Engineering and Management*, Prentice Hall, Englewood Cliffs, NJ, 1971, pp. 729–732, 746–748.

Eginton, W.W., Quality control as an integral part of the corporate team for product safety, in *Proc. of the 1976 American Society for Quality Control Conference*, American Society for Quality Control, Milwaukee, WI, 1976, p. 83.

Forman, H.I., Laboratory accreditation and its relation to product liability, in *Proc. of the 1978 Product Liability Prevention Conference*, p. 65.

Forman, H.I., Quality control, accredited laboratories, and product liability, in *Proc. of the 1979 American Society for Quality Control Technical Conference*, American Society for Quality Control, Milwaukee, WI, 1979, p. 184.

Greco, F.J., Federal products liability legislation: records retention and destruction problems as examples of need, in *Proc. of the 35th American Society for Quality Control Midwest Conference*, American Society for Quality Control, Milwaukee, WI, 1980, p. 94.

Hayes, G.E., *Quality Assurance: Management and Technology*, Charger Productions, Capistrano Beach, CA, 1982, pp. 209–243, 249, 295–393.

Jacobs, R.M., Evolution of quality/reliability due to litigation, in *Proc. of the 1983 Reliability and Maintainability Symposium*, p. 123.

Jacobs, R.M. and Mihalasky, J., Practices and systems for product liability prevention, in *Proc. of the 1976 American Society for Quality Control Technical Conference*, American Society for Quality Control, Milwaukee, WI, 1976, p. 108.

Juran, J.M., *Quality Control Handbook*, McGraw-Hill, New York, 1988, pp. 12–13.

Peterson, G.P., Inspection: a case study, in *Proc. of the 35th American Society for Quality Control Midwest Conference,* American Society for Quality Control, Milwaukee, WI, 1980, pp. 1–4.

The Travelers Insurance Companies, *Management Guide to Product Quality and Safety,* Hartford, CN, 1973, p. 90.

Wilson, N., Receiving inspection, in *Proc. of the 35th American Society for Quality Control Midwest Conference,* American Society for Quality Control, Milwaukee, WI, 1980, p. 7.

8 Financial Administration and Financially Focused Quality Training

Introduction to Financially Focused Quality

Financially Focused Quality (FFQ) aims at getting all employees to think as if they were members of the finance community. By doing so, FFQ enables and empowers every employee to contribute to increased profitability. This is accomplished as follows:

1. Educate every employee on the financial information necessary to understand the following:
   a. How the company operates (for example, how it earns a profit)
   b. What the finance organization is and what it does
   c. How each employee contributes to the bottom line
   d. How other circumstances (suppliers, customers, other company employees) affect profitability
2. Present to every employee the FFQ methodology for analyzing potential process improvements and corrective actions.
3. Stimulate and encourage employees to utilize a Financially Focused Quality mindset when performing daily responsibilities and making decisions.

Financial Training

In order to utilize financial data, employees need to understand the world of finance. The most common method for gaining such an understanding is to embark on formal classes at institutions of higher education. In increasing numbers, working professionals are enrolling in universities designed specifically for adult higher education.

Accredited in 1978, the University of Phoenix (UOP) was among the first to recognize the need for degree and continuing education programs for adult professionals. In 1999, UOP had over 81,000 enrolled students, making it America's largest private accredited university for working adults. UOP is just one of many such institutions. Adults wishing to learn finance principles can take classes such as:

- Accounting (e.g., cost accounting, financial accounting)
- Cost estimating
- Business administration (e.g., management, marketing)
- Economics (e.g., micro-economics, macro-economics, the economics of state and local finance)
- Quantitative methods for business analysis
- Finance (e.g., personal finance, financial management, small business finance, managerial finance, financial theory, international finance, theory of corporate finance)
- Investments (e.g., introduction to investing, portfolio analysis, speculative markets, corporate financial investments)
- Principles of real estate (e.g., income property analysis, residential real estate, site location analysis, mortgage markets)
- Financial institutions (e.g., bank administration, commercial banking)
- Insurance (e.g., life, health, property, and liability insurance; risk management)
- Working capital management (e.g., capital budgeting)
- Business cycles and forecasting
- Money and capital markets
- Commercial bank management
- International financial management

These classes are geared for the adult with a full-time job; for example, the classes can be offered in five or six weekly 4-hour sessions, or in 6- or 8-hour weekend sessions. Such scheduling enables working adults to attend classes without interrupting their work schedules. The rapid growth in adult education providers is proof that this approach has been very successful.

The education offered by outside institutions is generic in that the classes pertain to a wide range of businesses, and it is possible that much of the classroom information would not be of value to those employees content with their employers and their positions.

A perfect implementation of Financially Focused Quality has every employee in the company receiving an education in the financial areas that are unique to that company. Seldom will an outside institution be approached to supply such education; however, another alternative exists. It is much more affordable and practical for the company to tailor a unique course aimed at the financial issues pertinent to the enterprise. This chapter presents the basic financial education concepts that apply to almost every company and is based on the financial concerns of a very large financial operation, but the principles are easily applied to smaller businesses. Some suggestions for training employees in company-unique financial topics appear at the end of this chapter.

Cycle of Financial Activities

Figure 8.1 presents an overview of financial activities for a large company.

- *Estimating* refers to the process of establishing prices for products and services. For many businesses, including contractors and service providers, this usually involves preparing, proposing, and negotiating contracts.
- *Budgeting* ties specific performing or responsible business functions to cost targets for accomplishing required tasks. For operations working under contracts, this function usually takes place after contract award, and budgets are established for tracking performance against the negotiated target costs.
- *General accounting* involves the routine tasks associated with a business (e.g., accounts payable, accounts receivable, etc.).
- *Reporting* provides management and (when appropriate) the customer with factual detail regarding performance to budgets, and at the same time offers forecasts and corrective action taken to resolve budgetary problems.

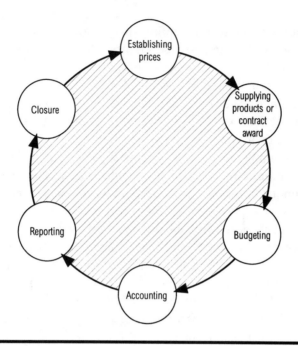

Figure 8.1. Cycle of Financial Activities

■ *Closure* is the activity that allows the enterprise to book a specific sale. This could be very simple, such as after a retail sale when the customer walks away with the product and the receipt. With a very large contracted sale, however, closure can become very cumbersome. Often, a contractor will not receive final payment for the work performed until a thorough closure activity is completed. For government contracting, this can involve the finalization of overhead rates and disallowances (discussed later).

This cycle begins when a product or service has been designed or planned. The Finance organization determines the price via a method known as estimating. Once the price has been established, orders are taken and contracts are initiated. For major, one-of-a-kind type services (for example, construction projects), the company prepares a proposal and submits it to the potential customer. The negotiation process is concluded when both parties sign a contract.

An important element of the contracted terms is the target cost to be paid by the customer to the company for performance of the agreed-to tasks. Based on orders and/or negotiated contracts, budgets are established based on a forecast of what is scheduled to be delivered. These budgets are tied to the original estimated cost. Actual costs are tracked periodically (for example, monthly). Tracking costs makes problem areas visible and enables timely corrective action.

When the budgets are established, a tracking system is in place, and the company is operating at full steam, it is essential that the accounting functions be performed smoothly. The major accounting functions are

- Payroll accounting
- Supplier accounts payable
- Financial accounting
- Cost accounting
- Accounts receivable
- Property administration

Now it is time for the reporting process to begin. In large companies, quite sophisticated financial systems can be employed to assist management in the tracking of actual performance against cost and schedule criteria. These activities continue until the last contracted piece of hardware or service is delivered, and then the contract closure process begins.

While the above financial functions are being performed, the company should be actively pursuing new business by investigating new products and services. The actual costs encountered in the process of manufacturing or preparing previous products should be used as a basis for establishing prices for new products and proposing new contracts.

Developing innovative products and preparing new proposals are vital to an ongoing business entity because such procedures increase the probability of future business, which in turn enables the cycle to continue.

Financial Concepts

As with any profession, those in Finance have their own unique set of concepts that are necessary to function successfully on the job. Comprehending the key financial concepts applicable to most businesses will enable employees to better exercise the Financially Focused Quality mindset. These concepts are

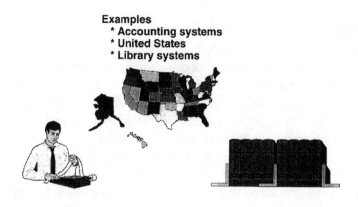

Figure 8.2. Examples of Classification

1. Accounting classifications
   a. Direct costs vs. indirect costs
   b. Labor costs vs. nonlabor costs
   c. Direct-direct vs. allocated-direct costs
   d. Work breakdown structure
   e. Cost accumulation structure
   f. Charge numbers
2. Timekeeping/cost segregation
3. Forecasting personnel

Accounting Classifications

Classifications provide detailed identity to large groups or divisions of information (see Figure 8.2). A reasonable person would not attempt to enter a major public library and begin to search for a specific book without making use of the card file. The U.S. Post Office utilizes Zip Codes to aid in sorting and delivering mail. Just as libraries and the Post Office make use of classification systems, so too does the world of Finance. Finance and accounting systems provide detailed identity to large groups of business expenses through classification

Direct vs. Indirect Costs

Costs may be classified as either direct or indirect (see Figure 8.3). Direct costs are expenditures directly benefiting and identifiable to projects. Direct costs may also be allocated direct costs, which provide specifically

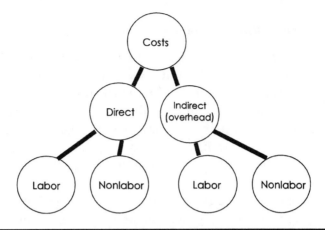

Figure 8.3. Accounting Classifications — Allocations

for the research, design, development, or production necessary to accomplish project requirements.

Indirect costs are those which, under normal requirements for expenditure, do not directly contribute to the accomplishment of project requirements. The term "indirect cost" is sometimes used synonymously with "overhead".

In some larger enterprises, it is not uncommon for employees to be referred to as direct or indirect employees. In this situation, direct employees are those who must account for every minute of their time as benefiting a specific contract charge or an overhead function. Indirect employees are generally required only to account for time spent on uncommon overhead functions (seminars, jury duty, etc.).

Companies that do business with the U.S. government are required to define the detailed make-up of their direct and indirect costs in their company disclosure statements, which must be approved by government representatives.

Labor vs. Nonlabor Costs

Costs also may be classified as labor or nonlabor. Labor costs are those paid to company employees. Such costs, at the lowest level, are generally expressed in hours or fractions thereof, to which direct labor rates are applied to obtain dollar values. Nonlabor costs are those paid to suppliers or subcontractors for purchased parts or services.

Direct-Direct and Allocated-Direct Costs

Direct costs are further classified as direct-direct or allocated-direct. Allocated-direct costs are still direct costs; however, they are not directly assigned to a specific project. They may benefit multiple contracts or multiple projects within a contract, or they may be incurred in a manner that makes identity to specific projects not economical — for example, they may be incurred through process-type operations. The terms "allocated prime cost" and "pool work order" have also been used for classifying allocated-direct costs.

An example of a process-type operation is mixing paint. The paint mixing process may take only a few minutes per container, and the paint could be used to support any number of products and/or projects. Such costs are accumulated in pools and are allocated to projects and contracts on some previously defined basis.

Overhead Costs (Indirect Costs)

Overhead costs also are accumulated in pools. These pools are allocated to contracts on some previously determined basis that may or may not be in proportion to the benefit gained. Similar to the case of Allocated Direct Costs, certain overhead expenses are not directly related to a specific overhead pool, but are accumulated separately and then redistributed (allocated) to benefiting pools. Industrial relations, industrial security, office operations, plant services, and material handling functions might exemplify such costs.

Work Breakdown Structure

When contract requirements dictate classification of costs by components of hardware and services or by organization, the customer is provided these costs by use of a work breakdown structure (see Figure 8.4), which is a product-oriented, family-tree division of efforts related to hardware and services that breaks the contract work down into ever smaller units.

A work breakdown structure is used for budgeting the work to be performed. The highest level of this structure is the total effort to be performed. Similar to the blueprints for a construction project, the work breakdown structure represents the major portions of the project for which performance measurement is deemed necessary. The assignment

Figure 8.4. Accounting Classification — Work Breakdown Structure (WBS)

of necessary reporting levels near the top of the structure is most often the responsibility of the government, while the lower levels are established at the option of the company to enable management visibility of expenditures and control of resources.

Cost Accumulation Structure

Many larger companies, and particularly those engaged in business relations with the U.S. government, use a cost accumulation structure (see Figure 8.5). This structure provides for collection of costs at the lowest level where company effort (resources) is expended. Costs are then summarized through successively higher levels of both the work breakdown structure and the organizational (functional) structure.

Charge Numbers

Many large enterprises use charge numbers in a time-recording system (for example, timecards) for assigning costs to specific projects or classifications. These charge numbers, sometimes referred to as work order/work authorities, are a numbering system that is the basis for summing costs into the accounting (and allocating) structure, the functional organization structure, and the work breakdown structure.

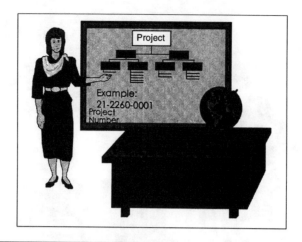

Figure 8.5. Accounting Classification — Cost Accumulation Structure

Timekeeping, Timecards, and Cost Segregation

A timekeeping system uses a character coding system to achieve required segregation of costs. This character coding is accomplished through the use of a charge number (see above). For direct employees, the charge number for the effort to be performed by the employee is entered on the timecard at the time the effort is started on that day. The number of hours worked on that particular job also is entered.

Timecards are essential to most employers requiring close scrutiny of employee time. They are the official company time records. They are the basis for paying employees and for obtaining reimbursement from the customer. Figure 8.6 presents an example of a 10-character charge number coding structure. In this particular figure, a 10-character coding system is used; however, the company has no restrictions as to the number of digits used in the charge number. Also, the system illustrated here is fairly complex, but the company may design its coding system as deemed appropriate.

In this example, the charge number is 42-BBX2-8411. Notice that each character of this number identifies specific costs and reads in two different ways depending upon the first digit. The first digit (sometimes referred to as the major class code) identifies whether costs are direct or indirect. In the example, the first digit is a 4, signifying that the charge is direct. Had this digit been a 1, 5, 6, 7, or 9, it would mean that the charge is indirect.

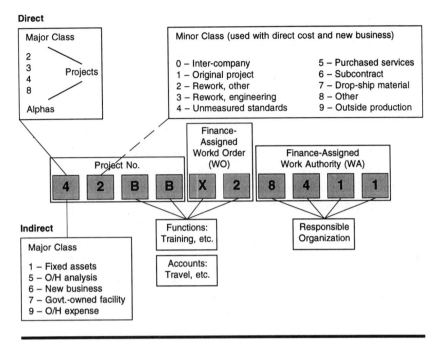

Figure 8.6. Example of a 10-Character Coding System

The minor class code (second digit) identifies the type of cost (for example, inter-company, original contract work, subcontract, etc.) for direct or new business type of activity. The number 2 here classifies the effort as "rework, other". In this example, the minor class code serves no purpose for indirect charges; therefore, company guidelines might require that a zero be used here for indirect work.

The next four digits may be referred to as the work order. For direct efforts, this is often where specific contract items are differentiated. In this example, the BBX2, when combined with the major class code 4, means that the effort is supporting item 2 of a particular contract. For indirect effort, the overhead function is designated here (e.g., training class, jury duty).

The last four digits are referred to here as the work authority. For direct efforts, this is where the specific tasks might be identified (e.g., receiving inspection, design review). In this example, the work authority is 8411, where the 84 means the effort is the responsibility of the Product Assurance department, and the 11 identifies the activity as being shipping inspection. For indirect activities, these four digits are used to identify the department whose overhead budget is funding the indirect activity (e.g., the department of the employee serving on jury duty). The charge number and its corresponding coding system might be considered the Zip Code of Finance.

Virtual and Manual Timecards and Labor Recording

Many companies utilize PC-based financial systems to account for effort by their employees. Others still use timecards. In many instances, charge numbers are entered into the system. It has been said for government contractors that, when charge numbers are used, correctly recording activity is the most important function that employees perform. Charge numbers identify labor expended for particular activities so that it is charged to the proper contract or overhead account. But, why should correct recording of time be the most important factor?

Accurate timecard records are necessary to obtain reimbursement from the government, as well as to establish wages and salaries to be paid to the employees. For any government contractor, misuse of the timecard (for example, knowingly charging to an incorrect charge number) is a serious offense analogous to forging a check; such an action can result in lawsuits and/or loss of government work for the company. The offending employee would be subject to disciplinary action, such as loss of position or employment and criminal penalties (i.e., jail, fines).

Figure 8.7 illustrates a simplified example of how the timekeeping method discussed above could be applied to record the work provided by a typical direct employee working for a government contractor. On Monday and Tuesday of the week, the employee works 8 hours each day on a specific Air Force contract. The major class code for direct activity is an alpha character, and in this case the "A" designates that this effort is being performed for an Air Force customer. The remainder of the charge number specifies the effort in finer detail.

Wednesday and Thursday, 8 hours are worked each day in support of a Navy contract (identified by a major class code of "N"). On Friday, the employee first spends 4 hours at a medical appointment (major class code 5 for overhead) and wraps up the day working on the same Air Force contract he worked on Monday and Tuesday.

Timecards can include other fields for data entry, such as overtime and special work hours. When the employee signs the timecard, he is in effect signing a legal document. He is saying that he has performed the

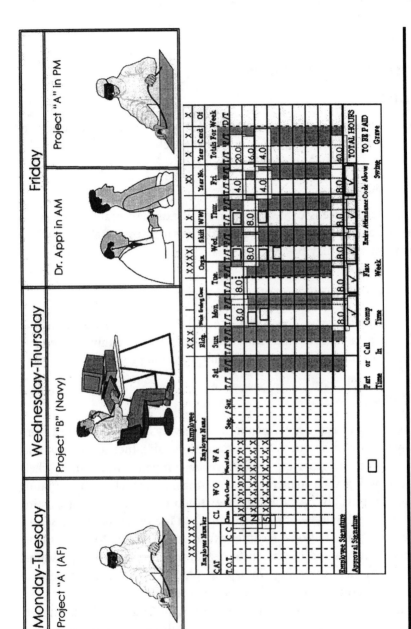

Figure 8.7. Timecard/Timekeeping

work as defined on his timecard, and in return the company will pay him for his efforts.

Forecasting Personnel

Criteria

The accurate forecasting of personnel is vital to financial success, as discussed for the following sampling of affected areas.

Proposals

A contractor needs contracts in order to exist. Accurately forecasting personnel requirements is essential for preparing proposals from which such contracts may be awarded.

Training

Just having bodies in the right place at the right time is not enough to fulfill contractual requirements. Such bodies must be fully trained and capable of making positive contributions to the product.

Facilities

Forecasting working space requirements is very important when a company is facing either increasing or decreasing sales. Increasing personnel usually means increasing workspace. A company would most likely accommodate such an increase by entering into leases for more building space or possibly by building new structures or modifying existing facilities. It should be obvious that prudent use of resources is essential to ensure that additional facilities expenses are limited to what is truly needed and wanted by the customer. Similarly, decreasing personnel usually means decreasing workspace. Not so obvious is the need to routinely scrutinize facility forecasts and requirements.

In the rapidly changing environment of the 21st century, the uncertainty of business trends and changing technology means that most companies should exercise extreme care when examining facility leases. Every time a lease is up for renewal, that lease should be examined as if it were a completely new lease. Consolidations of working areas should be achieved at every opportunity to ensure that each square foot of leased and owned space is necessary and fully utilized.

Human Resources

Human Resources must have accurate personnel forecasts in order to provide cost-effective operation of the following services:

- Employment
- Terminations
- Conflict of interest support
- Maintenance of policies and procedures
- Compensation analysis
- Benefits coordination
- Health services (e.g., medical and dental plan coordination)
- Medical services (e.g., in-plant doctors and nurses)
- Equal opportunity programs
- Labor relations
- Security/emergency services (e.g., in-plant fire department, ambulance)
- Government security (e.g., document control, security clearances)

Forward Pricing Rates/Overhead and General & Administrative Costs

Overhead and general and administrative (G&A) costs are often allocated to personnel, and forecasts are used to develop forward pricing rates.

Forecasting Direct Personnel

The average worker in the U.S. is paid for a 40-hour work week. This most frequently results from Monday-through-Friday employment of 8 hours each day. Saturday and Sunday are grouped together and are referred to as the weekend. In addition, most companies provide holiday compensation of some sort. Most often this takes the form of days off with pay, or extra pay for working during holiday periods (e.g., time-and-a-half, double-time, triple-time, etc.). The 8 hours that an employee is paid to work each day (excluding holidays) is called straight time, which is the starting point for determining the total time an employee will spend performing productive work.

Statistically "average"
direct employee

	Yield/Realization
Criteria	**Hours/day**
Straight time (S/T)	8.0
Vacation/sick leave	−0.7
S/T available for work	7.3
S/T work on overhead	−0.5
S/T available for projects	6.8
Overtime (O/T) @ 2%	+0.1
S/T + O/T available for contracts	6.9

Figure 8.8. Forecasting Personnel: Realization Factor

When forecasting personnel, management first examines the future work with regard to the following three areas:

1. *Backlog of work:* The activities that are scheduled to satisfy current orders or contractual requirements
2. *Forecast or proposed:* Effort to support areas where additional orders are anticipated, or for effort that is in the proposal phase
3. *Potential:* Activity to support new products that may be designed or new projects that may be proposed and negotiated and begun within the time span of the forecast

The forecasted hours associated with these three groupings are converted to estimated numbers of direct employees, called direct head count. This conversion is achieved by applying a yield or realization factor.

Yield/Realization Factor

This factor takes into consideration the effects that vacation/sick leave, work on overhead accounts, and overtime will have on the time that the statistically average employee will perform profit-generating work (see Figure 8.8). The calculation is as follows:

■ Straight time: 8.0 hours

For this illustration, a 40-hour, 5-day workweek is used, with the average employee working 8-hour days.

■   Vacation/sick leave: –0.7 hour

For each 8 hours of straight time, 0.7 hour is removed for vacation and sick leave. For the company in this example, there are 250 working days a year. 250 days times 0.7 hour equals 175 hours that are spent each year by the average employee taking vacation leave and sick leave. That 175 hours equates to just over 4 weeks a year. After making the vacation and sick leave adjustment, the average employee has 7.3 straight-time hours available for work each day.

■   Overhead accounts: –0.5 hour

The average employee does not spend all of this available time on activities that directly generate profits. He/she occasionally will spend time on overhead activities, such as the following:

1.   Engaging in independent research and development studies (the results of which potentially enhance the company's position in technical expertise)
2.   Working on bid and proposal activity (efforts to acquire new business)
3.   Attending training seminars or classes
4.   Taking care of union business or other appointments not related to contract work
5.   Participating in work-related professional societies
6.   Performing jury duty or appearing as a court witness
7.   Dealing with an act of God (for example, a natural disaster such as an earthquake) which makes it impossible or impractical for the employee to come to work
8.   Taking part in non-work related activities felt by the company to provide other benefits (e.g., blood bank, Junior Achievement)

In this example, 0.5 hour a day (125 hours a year) is the adjustment made to the 7.3 subtotal. Therefore, the 7.3 straight-time hours available in each 8-hour day are reduced by this 0.5 hour, yielding 6.8 straight-time hours remaining for every 8-hour working day. Does this mean that, on the average, an individual employed for 8-hour days (5 days a week) will contribute only 34 hours (5 × 6.8) for activity that directly generates profit? No, it does not. There is one more factor to consider.

■   Overtime: +0.1 hour

The average employee works a certain amount of overtime. In this example, the average overtime per 8-hour day is 0.1 hour. The overtime yield factor is based on the percent of overtime worked by direct employees on contract work; therefore, straight time (6.8) + overtime (0.1) available each day = 6.9 hours. The result of these computations is that the average employee working an 8-hour day can be expected to contribute 6.9 hours per day directly to contracts.

Forecasting Problem

A manager is responsible for delivering 690 direct hours per day on his various contracts. How many employees would he need to employ on these contracts? 100 direct employees would work an average of 690 contract direct hours each work day. This is calculated as follows: 6.9 hours × 100 employees = 690 contract direct hours.

Contracting Modes (Fixed Price/Cost Reimbursable)

There are basically two types of contracting modes associated with contracting: fixed price (FP) and cost reimbursable (CR). The modes of contracting are directly related to the amount of cost risk to the contractor. These modes are outlined below.

Fixed Price

This type of contracting is far and away the most common type. It implies that there is a set (fixed) price that the customer will pay for contracted effort, except as allowed by clauses in the contract. Fixed price is the mode often chosen when contracting for routine tasks, such as automobile maintenance or house painting. A profit margin should be included in the quoted price. The fixed price mode can be broken down further as either a firm fixed price (FFP) contract (most common) or a fixed priced incentive (FPI) contract.

Firm Fixed Price

A firm fixed price contract exposes the contractor to the greatest financial risk. Under an FFP contract, the company is obligated to perform until completion of the task, regardless of cost. Say, for example, that a contractor's quote for $20,000 was accepted and a contract was signed. The contractor may find later that an error was made in the quote and that the job will cost him more than $20,000, but he is contractually obligated to complete the job even though he will lose money.

Fixed Price Incentive

An FPI contract offers the possibility of the final price paid being higher or lower than the negotiated target cost. This results from pre-negotiated incentives. Incentives can be either positive or negative. With a cost incentive, if the company spends more than the negotiated target cost, the customer will pay a portion of the increased cost. On the other hand, if the contractor spends less than the contracted amount, the customer will pay less. Incentives also may be based on performance in regard to schedule and technical objectives.

Cost Reimbursable Contract

A cost reimbursable contract (CRC) provides the most protection, though often has the maximum technical risk. Assume that Motorola Company wants Lockheed Martin to design a missile that can launch satellites into orbit for a high-tech communications project. Lockheed Martin will be careful before entering such a contract in a firm fixed price mode, because potential problems could cause the expenses associated with such a project to skyrocket and result in tremendous losses. Unlike FFP contracting, where all contracted tasks must be performed, CRC performance is only required up to the limitation of the cost specified in the contract. This limitation is usually the target cost of the contract. The cost reimbursable mode can be broken down further as cost plus fixed fee, cost plus incentive fee, or cost plus award fee.

Cost Plus Fixed Fee

Cost plus fixed fee (CPFF) assures the company of a fixed fee, regardless of final cost outcomes.

Cost Plus Incentive Fee

Cost plus incentive fee (CPIF) increases or decreases fees based on cost, schedule, or technical performance.

Cost Plus Award Fee

Cost plus award fee (CPAF) increases fees based on a subjective evaluation of the final product performed by the customer's award fee team.

Disallowances

The typical employee working for a government contractor flies coach class to Washington, D.C. for company business. The government generally will reimburse the company for the costs of coach class travel when required. However, do you think the chief executive officer (CEO) of a major contractor flies coach class to Washington, D.C.? No way! You will find that CEOs almost always fly first class. The company pays these costs, but do you think the government will reimburse the customer for the full cost of the first-class tickets? No, again. The government will reimburse only for costs that are deemed necessary, and first class travel is seldom — if ever — considered necessary. As a result, such costs above what is essential are disallowed, which means they will not be reimbursed. A company chooses to fly executive officers first class knowing these costs will be disallowed.

Rates per Direct-Labor Hour

It is not unusual to find a posted shop rate at an automotive repair shop. Such a rate may be listed as anywhere from $30 to $90 an hour. This does not mean that the mechanic earns this high wage. The mechanic is earning less, and the difference between the posted rate and the rate charged by the repair shop is overhead and profit. Figure 8.9 illustrates this principle. While an hourly employee may be paid a gross wage of $21.55 each hour, the company is charging the customer much more.

An example of the components of the direct-labor rate follows:

- The employee earns a gross wage (before taxes) of $21.55 per hour.
- The average employee in this company also receives fringe benefits, such as vacation leave, sick pay, or a retirement plan, which equates to approximately another $9.46 an hour.
- Overhead must be included, as well. A portion of the overhead pays for indirect labor (e.g., managers, vice presidents). $10.34 is added to the labor rate for indirect labor.

	Dollars/ hour
Employee direct wage	21.55
Employee fringes	9.46
Indirect labor	10.34
Indirect fringes	4.54
Occupancy	3.88
New business	4.96
Other	8.41
Total cost	63.14
Fee or profit	6.31
Total dollar amount	69.45

Figure 8.9. Rates per Direct-Labor Hour

- The indirect employees also have fringe benefits, so another $4.54 is added.
- Costs associated with the company facilities are included (such as depreciation and maintenance), so $3.88 is added.
- The customers recognize that an ongoing business concern is always performing special projects in an effort to generate new business. A certain amount of these costs are anticipated, so $4.96 is added in this example.
- Other costs might include insurance, corporate management allocations, supplies, equipment rental, professional outside services, and indirect travel; here, a figure of $8.41 is used.
- An ongoing business concern should be making a profit for the stockholders, so a profit component of $6.31 can be tacked onto each direct labor hour. Thus, in this example, the total direct labor figure is $69.45 per hour.

Teaching Financially Focused Quality and Financial Concepts

Financially Focused Quality (FFQ) requires that all employees maintain a financial focus throughout their work day. This is best facilitated by offering employees financial training. Reading this text is an excellent means by which to grasp fundamental concepts and learn about FFQ tools. The next step in implementing FFQ is organization-wide education. In smaller companies, the mere reading of this text with interactive discussions can create the Financially Focused Quality mindset that fosters significant savings. Larger companies have taken this further by making formalized training available to all employees.

Of utmost importance when considering formalized FFQ training is to remember that resources should only be used for what is needed and wanted. In the early days of TQM, the executive boards of many companies would give a blank check to their training organizations to develop TQM courses. The result was hundreds of millions of dollars in time and resources expended to teach vast concepts, many of which were never used.

For example, a corporate executive was asked why his company had recently spent $500,000 to send 100 staff members to a TQM off-site training seminar. He responded, "We wanted to let the employees know that we are quite serious about this TQM." Unfortunately, attendees gave very little favorable feedback regarding the event. In the instance, management had not performed a process improvement cost analysis (see Chapter 10) before deciding to schedule the seminar. If they had, the company could have saved half a million dollars.

An Effective FFQ Training Guideline

Financially Focused Quality does not allow such an indiscriminate approach to training. A formal FFQ training program should have the following qualities:

1. FFQ training is available to all employees.
2. Company employees attend one session, which is 2 to 4 hours in duration, with two or three 10-minute breaks.
3. Each session should be taught as a seminar that allows interactive discussion about topics covered.
4. The seminar should be attended by no more than 80 students at a time, and then students should be from a similar discipline within the company. This ensures that seminar discussion and questions are relevant to all in attendance.
5. The class is formatted in four modules as follows:
    a. Module 1 — *Introduction to Finance* (30 – 60 minutes): A tailored introduction to the company's financial health and

organization structure, presented in the following format: (i) presentation of key financial data unique to the enterprise (for example, sales, earnings before interest and taxes, earnings per share, backlog); (ii) a description of how the company obtains and uses funds to generate a profit; (iii) summary of the financial organizations, with the names of managers and employees to call with questions.

b. Module II — *Financially Focused Quality (FFQ) Overview* (15 – 30 minutes): (i) Presentation of basic FFQ tools; (ii) use of case studies and examples of FFQ success stories to illustrate the application of FFQ tools. It is best if the case studies are drawn or developed from the sponsoring company's own FFQ experience, although examples from this text can also be used.

c. Module III — *Financial Concepts* (30 – 60 minutes): Not all the concepts presented in this text are applicable to every company; therefore, only relevant financial concepts should be included in the module.

d. Module IV — *Implementation* (15 – 60 minutes): The course concludes with a summary of recent company programs relevant to the attendees. This may include quality management issues (Chapters 2 and 3), such as certain employee benefits, recent restructuring or reorganizing activity, rightsizing, and outsourcing. Topics such as these will stimulate much discussion.

The facilitator of this seminar should take every opportunity to encourage the attendees to be financially focused. Employees should feel challenged to examine old and new work processes with a Financially Focused Quality mindset, asking such questions as:

1. Do I really need to be doing this?
2. Is there a way I can accomplish this more quickly and effectively, either using the latest technology or just plain old common sense?
3. Is there an activity that another employee or group is performing to support me that I really do not need or could be made easier?
4. Is there anything anywhere in this company that I feel could be improved?

Getting an accurate answer to the above questions will be possible after the employee has completed the first three modules. If the answer to any of the above questions is "yes", the appropriate FFQ tool should be initiated (for example, process improvement recommendations).

Examples of Financially Focused Training Materials

Some of the figures contained in this text were developed from presentations and handouts used for Financially Focused Quality training classes that have been taught to literally thousands of Lockheed Martin Missiles and Space employees. These classes have been taught to groups encompassing all disciplines. They have been tailored using unclassified Lockheed Martin financial statements available to the public (for example, annual reports to stockholders.) Classes average 1 to 4 hours. There have been classes designed specifically to teach Financially Focused Quality to such program-specific areas as fleet ballistic missile and government customers, as well as the employee population at large.

Self-Paced Training

Another method for teaching FFQ concepts is self-paced training. Technology has made this a very simple process, using already developed formalized training materials. Classroom charts can be formatted and placed on floppy disks that any employee can access inexpensively on a desktop computer. Mainframe-based, self-paced training is also a viable option. The next chapter presents several other methods (one-on-one training, staff meetings) for informal training and motivation of financially focused employees. All companies are encouraged to use a combination of such training techniques when implementing their own FFQ programs.

Summary

Understanding basic financial terminology and concepts is critical for achieving the Financially Focused Quality mindset. As the individual receives continued training in company-unique financial aspects, the tools of FFQ will offer the prospect of greater profitability.

Self-Study/Discussion Questions

1. If you were to design a company-unique Financially Focused training program for your company, what topics would you include? How does your company make a profit?
2. How does your company account for time?
3. Assuming a 6.9-hour per day yield factor, how many employees working for one week would be necessary to deliver 1725 hours?

Reference

Hearn, E., *Federal Acquisition and Contract Management*, Hearn Associates, Los Altos, CA, 1990.

9 Financial Functions

Introduction

When most people think of employees in the Finance organization, they still think of accountants wearing green eyeshades pouring over 80-column worksheets. Nothing could be further from the truth! The advent of personal desktop computing makes possible the widespread communication of financial data at the push of a button and with much improved accuracy. As a result, more and more employees have access to data which enhances the opportunity to implement Financially Focused Quality (FFQ). This chapter explains the financial activities performed in many enterprises.

Proposals and Pricing

Every company needs to determine the prices at which it sells its products or services. Many of the procedures that apply to proposing new contracts can also be applied to the process of establishing prices. In either case, there is usually a historical basis by which to forecast costs associated with new products or projects.

Commercial Contracting

If a homeowner were planning to build an addition to his house, a routine similar to the following might be used:

1. The homeowner will generate a list of potential contractors. To accomplish this, he may (a) consult with friends and associates to obtain information about their similar experiences and to obtain referrals; (b) check the telephone directory Yellow Pages and various other advertisements; or (c) call the Better Business Bureau for recommendations and to access their records regarding the list of potential contractors.
2. With the list in hand, the homeowner will begin scheduling appointments with potential contractors, at which time he will provide the following to each contractor: (a) specifications (floor plans/blueprints/etc.) of the planned addition; (b) desired construction schedule; and (c) other preferred terms of the contract.
3. At these meetings, the homeowner will also want to discuss the contractor's performance history (including a review of past jobs, a list of references, etc.).
4. At the end of each appointment, the homeowner should have a feel for the contractor's qualifications and potential employability. As such, a request may be made to the contractor to provide a quote containing the following: (a) definition of work to be performed (special materials should be itemized); (b) proposed schedule (time frame in which work is to be performed); (c) complete cost; and (d) any other terms or conditions specified during the appointment (such as terms of payment, eventual customer selection of accessories, etc.).

By following a routine such as that outlined above, the homeowner would receive quotes that would provide the necessary data for narrowing down the field, eventually providing a sound basis for selection of a contractor for the job. The process whereby a contractor prepares a quote is similar to the government contracting process of preparing a proposal, as discussed below.

Government Contracting

In most government procurements, the customer often is not fortunate enough to have several options when choosing a contractor for many of the projects (e.g., missiles). This is because the expertise to develop and manufacture sophisticated equipment is understandably in limited supply. Therefore, in order for the U.S. to maximize and continuously improve upon existing technology, there is often one prime contractor in the best position for cost-effective performance. With only one legitimate contractor capable of developing and manufacturing a desired item, the company is designated a sole source. In sole source contracting, the primary goal of negotiations is a fair price for a safe, reliable product.

There are basically three types of proposals that the contractor uses to pursue government contracts:

1. *New business/solicited:* The government asks the company to develop and manufacture a product within certain specifications. For example, the Navy determines that the U.S. needs an antiballistic missile system with a range and accuracy far superior to any previously built. Desired specifications are generated, the contractor contacted, and, via the acquisition process discussed below, a contract eventually results that might call for the development of the system and delivery of 16 such missiles.
2. *Follow-on:* This second type of contracting calls for the procurement of more of the same product which has already been fully developed and purchased in a previous fiscal year. In regard to the new business/solicited example above, a follow-on contract might call for construction of 16 more of the same antiballistic missiles in the next fiscal year.
3. *New business/unsolicited:* The contractor asks the government customer if they would be interested in a new product with certain specifications.

A proposal is required for all three of the above procurement types.

Proposal Preparation Flow

The proposal preparation flow begins when the proposal leader issues a planning document, which performs the following functions:

1. Defines the proposal requirements
2. Schedules significant events requiring accomplishment prior to contract award
3. Contains the agreements or concessions reached by the customer and contractor representatives

When a requirement for a proposal is established, the Finance organization prepares and distributes quoting instructions to all involved company organizations, including the following:

1. Program office
2. Contracts (legal) organization
3. Finance organization
4. Operating branches/divisions/departments (the groups that will actually be performing most of the profit-generating effort directly related to delivery of the product)

Raw Resource

The Operating branches are the organizations that prepare the initial estimate of costs at the raw resource (lowest) level. For a manufacturing process, the raw resource is hours, to which labor and overhead rates are applied. The nonlabor (e.g., subcontract) raw resource is dollars, to which only an overhead rate (e.g., procurement burden) is applied.

Proposal Components

In addition to the estimated cost, the proposal includes the following three sections:

1. *Fee (cost reimbursable contract, CRC)/Profit (fixed price, FP):* Incentive/profit components not included in the estimated cost
2. *Terms and conditions:* Validity period of proposal, economic price adjustment clauses, warranty information, deliverables, period of performance, insurance, liability
3. *Basis of estimate:* Justification for raw resource requirements (discussed in detail below)

Basis of Estimate

The basis of estimate is justification by the contractor for the price quoted for a project. It states in raw resources the lowest level at which costs can be identified. The basis of estimate can be one of four primary types:

1. *Engineering estimates:* A staff of engineers performs analyses of what steps will be required to accomplish each proposed task.

2. *The same as in previous projects:* This method is used when quoting a task that has been quoted and/or negotiated previously. For example, the customer may want to order four more cable assemblies. These cable assemblies are identical to some for which a price was established in an earlier procurement. In such an example, a similar unit cost would be quoted. The proposal analyst often factors in a slight deviation in unit cost for such reasons as:

   a. *Learning curve:* The personnel performing this task have become more adept at its performance and, as such, can accomplish it more efficiently.

   b. *Rate factor:* An increase or decrease in the number of units being procured may sometimes justify a change in cost because of the economies of scale. Favorable economies may exist with increasing quantities when assembly-line approaches can be utilized, and suppliers offer discounts for ordering greater quantities of components. Conversely, unfavorable economies exist when quantities decrease.

3. *Similar to previous projects (with complexity factors/normalizing):* For example, the customer may want to order more cable assemblies but with a different connector. A previous proposal is used as the starting point, from which adjustments are made to tailor the quote to the new requirement.

4. *Percentage of effort:* Some tasks will relate so closely to other activities that the appropriate method will be percentage of effort. For example, the inspection of certain manufacturing processes historically may be at a level of 60% of the manufacturing hours (e.g., 10 manufacturing hours require 6 hours of inspection). A listing of such percentage-of-effort relationships includes:

   a. Inspection hours contrasted with total manufacturing hours
   b. Tooling hours contrasted with total manufacturing hours
   c. Computer-assisted design and documentation hours contrasted with total design hours

Financially Focused Quality, while encouraging short cuts and performing as efficiently as possible, discourages short-cutting in the estimating process, when the potential exists for underestimating. Using a percentage of manufacturing hours for deriving inspection resource requirements does not always consider the potential for failures, which could change inspection requirements without impacting those of manufacturing (e.g., test equipment failures). Similarly, tooling hours will not necessarily remain a constant factor of manufacturing hours. It has been said that he who takes a short cut sometimes takes twice as long.

Elements of a Historical Basis of Estimate

There are three basic requirements for utilization of an historical basis of estimate:

1. Identify scope of effort and the specific task quoted.
2. Identify historical source programs:
   a. Develop audit trail.
   b. Disclose documents, tapes, etc.
   c. Note date of documents, page numbers, work order/work authorities, and actual hours.
3. Establish relationship of historical programs to new programs:
   a. Identify program or contract.
   b. Develop complexity factors.
   c. Discuss the similarities.
   d. Explain why history is applicable.
   e. Provide rationale for factors.

Engineering Estimates

There are four basic requirements for utilization of engineering estimates to support resource proposals:

1. Disclose similar programs.
2. Explain reasons for not using historical data.
3. State that the estimate is based on judgment only.
4. Break the task down to the lowest practical level of detail.

Security Guidelines for Cost Proposals

Classified information is data not available to the general public. Classifications include sensitive, secret, and top secret. Government contractors have very specific procedures for handling situations in which classified data may be utilized. A general review of such procedures is contained in the following guidelines:

1. Data used in the basis of estimates should be kept unclassified.
2. When classified historical data are essential, use of such data must be (a) coordinated in advance with high-level management in the contractor's Finance organization, and (b) cleared by the appropriate program contracting officer (customer).
3. The estimate should include the statement: "An estimate based upon classified historical information which can be verified through recognized security channels and appropriately cleared Government Contractor Audit Agency personnel."

Audit Checklist for Basis of Estimate

A basis of estimate, which satisfies the following audit checklist, should comply with Public Law 87-653 and federal acquisition regulations:

1. Verify that the scope of work or task description is accurate in comparison to the work breakdown structure dictionary or to the proposal statement of work.
2. Ensure that the labor justification worksheets conform to accepted formats (should be specified in company procedures).
3. If at all possible, historical data must be utilized. If so, the following should be provided:
    a. Project or document name of historical program(s)
    b. Work order/work authority of historical program(s)
    c. Document number indicating where historical data are located
    d. Date of document
    e. Page number of document
4. If complexity factors or percentages are used:
    a. Factors or percentages must be supported with factual data.
    b. A description of the analysis performed and logic for the decision to use the approach should be provided.
5. Where engineering judgments are used:
    a. All potential problems should be investigated for similarities.
    b. Explanation must be provided as to why historical data from a similar program were not used as comparative data.

   c. A rationale must be developed that can easily be followed by an auditor.

   d. Estimated raw resources must be related to the program schedule.

   e. Standard bidding conventions must be used.

   f. The task must be described at the lowest practical level.

One of the most important aspects of the basis for the estimate process is ensuring high quality and accuracy of the data contained therein. For example, if math errors exist, the entire basis of estimate lacks credibility. The customer may award a very high score to a technical proposal but find the cost proposal deficient because of the lack of rationale for the cost. This deficiency could lower overall proposal scoring and contribute to a loss.

Pricing

As discussed earlier, the primary goal in negotiations is to arrive at a fair and equitable price for the contracted effort. The technique of pricing is used to convert raw resources into the total cost price. The rates applied to raw resources in pricing are known as forward pricing rates. A forward pricing rate agreement is usually negotiated separately with the customer on an annual basis. This saves the costs of negotiating overhead on each individual contract. A partial listing of forward pricing rate possibilities follows:

1. Direct employee labor salary/wage
2. Overhead
3. General and administrative
4. Allocated direct costs/computer service center costs (may include such costs as common quality services, common minor material, and computer-related expenses)

Budgeting

The budgeting process generally begins after a contract has been awarded. However, from time to time, a go-ahead may be given prior to finalization of the contract terms. Uncertainty regarding forward pricing rates, escalation factors, specific contract requirements, or a myriad of other situations could result in the lack of a signed agreement. If the formal contract has not been signed, budgets are established on an interim basis. Budgets can be definitive once a valid agreement is reached.

Before funding is allocated to performing organizations, funding restraints such as those listed below are considered:

1. Funding fences
2. Management reserve
3. Undistributed budget

Government Appropriations and Funding Fences

Government appropriations are divided into three major categories:

1. *Research and development (R&D):* R&D appropriations cover many phases of activities, from basic research to stopping just short of full-scale production.
2. *Investment:* Investment appropriations are commonly referred to as procurement funds.
3. *Operations:* Operating funds are open for 1 year, while R&D funds are open for 2 years. Investment or procurement funds have different time frames depending on the times being procured.

The contractor cannot move funds between appropriations, a limitation referred to as a funding fence. The term "different color of money" is sometimes used to refer to the different appropriations under which effort is negotiated.

Management Reserve and Undistributed Budget

Before funding is distributed (i.e., made available) for the performance of work, the Finance organization, in conjunction with program management, determines the funding to be set aside as either management reserve or undistributed budget. Management reserve is that portion of the total contract budget that is withheld by the contractor (i.e., not distributed) for management control purposes. Contractors normally withhold the management reserve for the following two reasons:

1.   To motivate managers to do the job at a lower cost than negotiated — instead of distributing the entire budget for the contract work authorized, a certain amount may be withheld as a management reserve. Management's effectiveness is evaluated in part by their performance to budgets. As a result, management diligently tries to accomplish their tasks within the budgets distributed to them. The withholding of a management reserve in this sense provides an incentive for management to reduce expenditures.

2.   To bank a contingency fund — a management reserve also provides budgeting goals for unanticipated program requirements that will impact future effort. Historically, most government contractors can determine for each contract the cost of problems and other program requirements that were unknown at the time of contract award. Using this as a valid history, after each new contract is negotiated, an amount of that contract value may be withheld from distribution. In this sense, a management reserve represents an amount of budget that the contractor feels eventually will be needed before contract completion, but does not know on what it will be spent.

An undistributed budget is a budget that is applicable to specific contractual effort but has not yet been assigned to a work breakdown structure element at or below the lowest level of reporting to the government. For the period of time that this effort remains undefined at a reportable level, it is designated as undistributed budget. Once the effort is defined at a reportable level, budget is distributed to the organizations responsible for its performance.

After the management reserve and undistributed budget are set aside, the remaining funds are distributed to the organizations responsible for performance of contract work. This process is often accomplished via a funding allocation plan that is similar to a contract between program management and the organizations responsible for performing the work. Each organization negotiates funding with program management, and eventually a signed agreement is achieved.

Reporting

The organizations that have signed the agreement and accept the funding are required to monitor their performance on a periodic basis and provide status reports to program management. These reports contain the following information:

1. Variances between budgeted costs and actual costs for the report period
2. Schedule variances
3. Reasons for variances
4. Management action taken to correct the variances (if necessary)
5. Variances forecasted at completion of effort

By alerting program management to potential problems, steps may be taken to correct the situation (e.g., the management reserve may be designated to cover over-budget situations).

General Accounting

Basic general accounting functions are necessary for most companies. Specific details vary from company to company. Provided below is a brief explanation of general accounting activities. The primary functions of general accounting are described as being central because they must be applied throughout the company, regardless of customer or product line. Such central accounting functions are described below.

Payroll Accounting

Many employees feel that the payroll accounting organization is the most important of all. Could this be because payroll accounting is responsible for supplying the periodic paychecks? You bet it is! Many employees don't give a hoot about any financial service except payroll. They want accurate paychecks delivered on time.

Functions of a typical payroll department include the following:

1. Collecting employee time charges, which can be done in two ways:
   a. *Timecards:* A common method for employees to record their time is with a timecard (either manual or virtual, as described below). Manual timecards usually need to be picked up and delivered to keypunch operators, who input

timecard data. Depending upon the size of the company, this can be a tremendous activity in itself. Where employees are paid on a weekly basis, timecards are usually picked up at the end of each pay period and delivered to individuals responsible for inputting the data to the payroll system.

    b. *Virtual timecards:* In the late 1990s, many companies began utilizing computers or telephones to input time charges. These paperless timecard methodologies are fairly expensive to implement, but studies have shown that large companies can benefit from such systems. Advantages of virtual timecards include (i) automatic electronic auditing against a master file, ensuring that accurate charge numbers are used to record time; and (ii) data no longer being keypunched into the payroll computer system (as there is now a direct telephone-to-payroll interface), thus eliminating the potential for keypunch errors.

2. *External deductions:* This involves the accurate accounting of such statutory and employee-authorized deductions as federal, state, and local taxes.

3. *Internal deductions/external deductions:* Many companies offer their employees the option of such internal payroll deductions as savings plans, savings bonds, and direct deposit to credit unions.

Accounts Payable

Accounts payable maintains an accounting system for the payment of liabilities and the recording of payment distributions. Generally speaking, the government will not pay a major contractor until the suppliers have been paid. As a result, a company is motivated to pay such suppliers in a timely fashion.

In fixed-priced contracting, progress payments are utilized to allow a contractor to be repaid for costs incurred prior to delivery of the product. Such payments tend to influence inventory levels negatively. Because progress payments are based on actual costs incurred, a company may procure more inventory than is needed because they know they will be reimbursed in a timely manner. Those companies which make progress payments to subcontractors are encouraged to utilize some sort of mechanism (e.g., audit) to ensure that subcontractors do not overbuy.

Financial Accounting

Financial Accounting performs the following tasks:

1. Maintains the general books of account
2. Maintains the subsidiary ledgers
3. Coordinates, prepares, and publishes company financial statements
4. Establishes and maintains control over all company billings
5. Coordinates billing policies with customers
6. Controls collection and disbursement of all company funds
7. Administers the company's travel reimbursement policies

Cost Accounting

Cost Accounting functions include:

1. Ensuring adequate segregation and accurate recording of direct and indirect costs
2. Maintaining a system of charge numbers and accounts and being responsible for the company's policies on classification of costs as direct or indirect
3. Establishing and maintaining timekeeping policies and procedures
4. Providing timekeeping guidance to operating organizations and monitoring conformance
5. Allocating overhead costs to products and services
6. Maintaining an internal audit group to verify that costs being billed to the government in overhead for cost reimbursable contracts are allowable costs, meaning that they meet the government's criteria for reimbursable costs (see Disallowances section in Chapter 8)
7. Assisting in overhead negotiations, including preparation of proposals and audit rebuttals
8. Maintaining financial control of receipt, storage, and disbursement of all company inventories

Contract Closure

When a contract is physically completed, it is placed in closure, and sequential steps (see Figure 9.1 and list below) are followed through to the final invoice. The time period can be as short as 1 year for fixed price contracts, or as long as 10 years for large cost reimbursable contracts. The average is 5 to 6 years.

1. The contracts organization formally notifies Finance that the last contractually required piece of hardware has been delivered.
2. Finance and the customer review financial cost data to determine allowable costs for cost reimbursable contracts.
3. Final overhead rates are incorporated.
4. Actual costs are reconciled with billed amounts.
5. Final fee proposal is prepared.
6. Upon notification of receipt of final payment, all financial documents are formally closed out.

All financial data related to this contract are readily available for supporting future contract proposals.

Summary

There is much overlapping in the cycle of financial activities. For example, budgeting is very much a part of the reporting process, and reporting information often is used in support of new proposals. Basically, financial operations can be viewed as providing four primary services:

1. *Accounting information:* This service makes visible the financial status of the company, including such areas as financial accounting, accounts payable, cost accounting, and property accounting.
2. *Contract pricing and cost control:* This area includes the following functions: (a) developing prices for products and services, (b) promulgating budgets, and (c) performing cost monitoring (comparison to budgets).
3. *Financial forecasting and analysis, and overhead and fixed asset budgeting:* This function provides insight into the changing

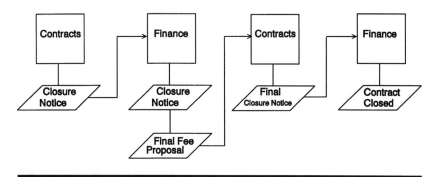

Figure 9.1. Contract Closure

business environment and allows effective planning and control of overhead expenditures.

4. *System of internal accounting controls:* Maintenance and performance of the internal accounting control system is a prerequisite for the safeguarding of assets and the reliability of financial records. It provides financial information to company and corporate management and to government agencies on behalf of the company, and financial support for the pursuit of business.

Note that Finance has access to all financial data. This access to information makes the Finance organization a natural centerpiece for activities designed to reduce costs and increase cost effectiveness.

Self-Study/Discussion Questions

1. Imagine that you are preparing a quote for installation of an air conditioning system in a two-story office building. What would a one-page summary look like? What raw resources would you use? What basis could you use to justify your estimate?

2. While on a business trip, you think of a way that your company could save significant travel dollars. To which of your company's organizations would you present your idea?

10 Financially Focused Quality Performers and Components

Introduction

The performers and components necessary for operating a large-scale Financially Focused Quality (FFQ) system are presented in the next two chapters. This chapter presents an FFQ overview and then focuses on the first phase of the process improvement activity: determining areas for improvement.

Financially Focused Quality enters the scene when either an opportunity for improvement is observed or there is some sort of failure. Therefore, all processes, procedures, products, or product components that are targeted for improvement or corrective action are appropriate FFQ subjects. The following is a sampling of the types of situations in which FFQ has been successfully applied:

1. Process improvement to streamline obtaining quotes from vendors in the advertising field
2. Developing a procedure for operation of hotel courtesy car service to the airport
3. Manufacture of an innovative product called "Penguin in a Helmet"
4. Training employees in the operation of a new MRP2 system
5. Identifying the ideal level of test console repair and maintenance support

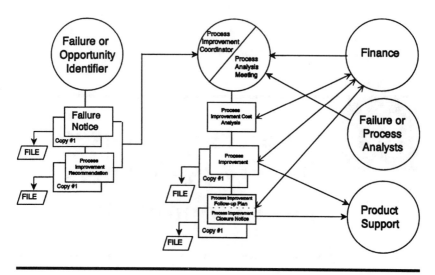

Figure 10.1. Financially Focused Quality Overview

6. In retail sales, job instructions for handling returned merchandise
7. In regard to manufacturer warranty service, making improvements in after-market support
8. Settling domestic disputes ranging from child-rearing to improvements in landscaping and interior design

The use of FFQ is particularly encouraged when considering such process improvement programs as continuous quality improvement, process management, employee suggestion systems, cost reduction or avoidance programs, Total Quality Management, total quality service, and value engineering. Imagine the effect a structured FFQ approach could have on the salaries, pensions, or benefits of corporate executives and government appointees. FFQ applications exist for manufactured products which have caused injury to users, as well as for the administration of nonprofit and charitable organizations. The list of potential FFQ applications is seemingly endless.

Financially Focused Quality Overview

As shown in Figure 10.1, there are 12 basic FFQ components. Few FFQ applications would require the simultaneous involvement of all of them. Figure 10.1 illustrates each component and its relationship to the others.

Failure Identifier or Opportunity Identifier

The Failure or Opportunity Identifier initiates the FFQ process by noting either a failure or a process for potential improvement. The process technically begins with completion of the appropriate document — Failure Notice or process improvement recommendation.

Failure Notice

A Failure Notice (FN) is used to document formal failures determined while specific job instructions were being followed.

Process Improvement Recommendation

The Process Improvement Recommendation is utilized when an employee observes an area or process for potential improvement. The proper form is forwarded to the Process Improvement Coordinator. The identifier makes a copy of the form and files it for future reference.

Process Improvement Coordinator

Choosing the individual to be designated as the Process Improvement Coordinator varies greatly, depending upon the form itself (either notice or recommendation). The coordinator might be the technical expert in the area involving the Failure Notice or the manager of the individual submitting the recommendation (see Chapter 11). The form is received and logged in. The coordinator then works with others to generate a listing of potential failure or process analysts.

Failure Analyst or Process Analyst

Failure or process analysts are divided into two groups:

1. *Potential failure or process analysts:* Individuals with even the slightest chance of being able to provide technical or administrative expertise in evaluating and/or resolving the reported situation. Included in this category are the product support organizations whose activities may influence the situation noted on the recommendation or Failure Notice.

2. *Primary failure or process analysts:* Individuals judged to have the greatest likelihood of positively influencing the process improvement or corrective action process. Again, the category may include the product support group influencing the noted situation. Primary analysts are invited to the Process Analysis Meetings.

Process Analysis Meeting

The Process Analysis Meeting is held so that the experts can determine what actions need to be taken. If the coordinator has not had FFQ training, a properly schooled representative from the Finance organization should also attend.

Finance

Finance is preferably represented by a coordinator who has received basic financial training via FFQ. As such, Finance is actively involved in the determination and evaluation of potential process improvements, the generation of process improvement cost analyses (see Chapter 11), the selection of the process improvement, development of the process improvement follow-up plan, and sign-off on the Process Improvement Closure Notice.

Process Improvement Cost Analysis

A cost analysis is generated for each potential Process Improvement.

Process Improvement

The final Process Improvement is selected and implemented. Finance tracks expenditures and ensures that costs associated with the final Process Improvement are in line with the cost analysis.

Process Improvement Follow-Up Plan

A follow-up plan is developed to make sure that Process Improvement is successful. Finance provides prompts to ensure adherence to the follow-up plan. The selected Process Improvement is forwarded to the product support organizations that will be performing the process.

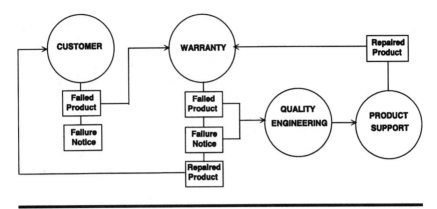

Figure 10.2. Failure Identifier (Warranty Service)

Product Support

Product support organizations implement the Process Improvements.

Process Improvement Closure Notice

The Closure Notice is forward to product support organizations when required. Finance ensures that costs are in line with the final instructions of the Closure Notice. All peripheral activities related to the implemented Process Improvement cease (for example, tracking unique costs).

Each of these basic components may have numerous subcomponents, which we will discuss below.

Failure or Opportunity Identifier

The application of Financially Focused Quality begins with identification of a failure or other opportunity for correction or improvement. Figure 10.2 presents the basic functions of a Failure Identifier. In this figure, the initial Failure Identifier is the customer, and the company Failure Identifier is an employee in the Warranty Service department. The Process Improvement Coordinator is a member of the Quality Engineering department.

1. The customer returns the failed product (e.g., television) to the Warranty Service department.

2. The Warranty Service department originates the Failure Notice (FN) and forwards it with the failed product to the Process Improvement Coordinator (Quality Engineering).
3. The Process Improvement Coordinator works with the product support organizations (e.g., Shipping, Manufacturing, etc.) and corrects the cause of the failure.
4. The product is repaired.
5. The product is returned to the Warranty Service department.
6. The Warranty Service department returns the product to the customer.

Anyone can be a Failure or Opportunity Identifier. For example, a dissatisfied customer, if given the opportunity to explain his discontent, is a *Failure Identifier*. A person realizing a different tool would help expedite the work is an *Opportunity Identifier*. Within the context of Financially Focused Quality, the individual who initiates the FFQ cycle by completing the appropriate paperwork is the Failure Identifier. In a company, that person should be the first person to identify a situation for potential improvement. Such conditions are likely to be determined by people in the following environments:

1. Routine failure-reporting mechanisms within business concerns, including:
   a. Any/all employees with customer contact, such as the Customer Service department, Warranty Service department, contract negotiators, or program office
   b. Quality Assurance inspection/test
   c. Employees given responsibility for reporting on performance to budget quality and schedule criteria
2. Informal failure reporting mechanisms within business concerns
3. Basically, any observations made by employees in the line of work, such as:
   a. Manufacturing employees experiencing difficulty in performance of assigned tasks
   b. Accountants who feel there must be a better way to deal with certain reporting requirements
   c. Supervision and management receiving comments from employees

Everyday citizens can identify failures and opportunities, and utilize FFQ tools and techniques to enhance decision-making and problem-solving in such areas as:

- Family disagreements involving a major purchase (e.g., car, home, appliance)
- Major personal decisions (e.g., high school graduate considering college choices)

Process Improvement Begins With Opportunity Identification

From discussion in the previous section, one can conclude that it is plausible that every company employee could be a Failure or Opportunity Identifier. However, the identifier designation is earned only when that person completes the required FFQ action(s). Many employees may know of a specific area where there is the likelihood for improvement, but only that employee taking FFQ action is recognized.

In most companies, the individuals or groups with primary responsibility for reporting failures will be obvious: customer service and inspection personnel. However, as all employees are potential Opportunity Identifiers, management needs to ensure that proper motivation exists. This section looks specifically at motivating employees to maintain a Financially Focused Quality mindset and to participate in the Process Improvement Recommendation process.

Methods for Failure and Opportunity Identification

There are many ways to identify an opportunity or failure, including the utilization of sophisticated inspection and test equipment. There is also the less sophisticated but often very effective method of using the human senses:

- *Sight,* for measuring and seeing that something does not conform to specifications, or for observing activities and processes. Also, the eyes are used to read about new methods, better procedures, and/or state-of-the-art equipment, which could improve productivity.

- *Hearing,* for hearing an unusual sound coming from machinery, listening to a customer complaint, or hearing a lecture, presentation, or discussion on new methods, better procedures, and/or state-of-the-art equipment which could improve productivity.
- *Smell,* for detecting an unusual odor or the presence of smoke.
- *Taste,* for determining, for example, if something is too sweet or too sour.
- *Touch,* for alerting someone to a slippery floor or a shock from a frayed electric cord.

Anyone recognizing a situation for potential improvement uses at least one of the five senses to make this identification.

The Failure and Opportunity Identification Process

Anyone who *thinks* a situation exists that might be improved upon could be an Opportunity or Failure Identifier — a shipping inspection employee who observes a dent in the product, the ultimate consumer, a mail courier who notices a truck with excess emissions while he is delivering mail. The word "thinks" is emphasized in this context, because for any number of reasons a perceived failure/discrepancy/opportunity for improvement may not, in reality, exist. Listed below are five situations in which a failure may be reported, when in reality the product (while in the factory) met the requirements.

1. The test equipment used to test a unit was not working properly. The unit met specifications, but faulty test equipment caused rejection.
2. The Failure Identifier (e.g., a customer) was not fully aware of the contracted specifications of the product.
3. The Failure Identifier (e.g., an inspector in the manufacturing process) misunderstood inspection criteria and rejected an acceptable unit.
4. The product was damaged in shipment from the factory.
5. The Failure Identifier (a customer) was not operating the product properly: (a) Even though the customer had the education

level of the target consumer for the product and read the instructions, he was not able to understand them, or (b) the instructions were clearly written, understandable, and easily accessible within the product packaging, but the customer did not bother to read them.

The company should recognize, though, that there was a failure of some sort in all but one of the above situations. Such failures are not in the design of the product, but could be due to the following reasons instead:

1. Test equipment failed.
2. Customer had false expectations, and there was a failure to clearly communicate details about the product.
3. Inspector misunderstood manufacturing criteria — management had failed to train the inspector properly.
4. Product was damaged in shipping due to a failure to package and/or ship the product properly.
5. Either the average customer was unable to comprehend instructions because the writer of the instructions failed to prepare clear operating instructions, or the customer simply failed to read the instructions.

The last failure mentioned is the only one for which the company would not need to consider a Process Improvement. In this case, the customer should have at least tried to read the instructions.

When a sales contract is prepared or a sale is made, an intellect level slightly below average must be assumed for the prospective customers; however, analysis may reveal that certain limits must be placed on the extent to which a company goes when packaging and selling its product. For example, a company that manufactures small model toys is aware that an infant could potentially choke on a toy; therefore, the following steps could be taken to reduce the chances of such an event occurring:

1. The package is clearly marked with the statement "Not recommended for children under the age of 4."
2. Instructions included with the toy state: "Caution — Toy should be kept out of reach of infants, and should not be placed in mouth."
3. A mechanism is rigged inside the toy package to play a taped message that says in a loud booming voice: "Caution! Toy

should be kept out of reach of infants and should not be placed in mouths."

4. The company stations security officers at the cashiers everywhere the toy is sold. Before a sale is allowed, the officer explains the hazards of the product, and the customer must sign a declaration that he or she understands the risks.

In determining how far a company should go to ensure clear communications with the customer, one must consider what is *reasonable* (see Chapter 6). For the sake of this text, a failure shall be defined as a discrepancy or defect that is determined by a reasonable person. This definition is used regardless of the Failure Identifier, and it must be assumed that the identifier has, at best, intelligence that is slightly below average, and that this person is also reasonable.

Customers

There is no doubt that, after a product is manufactured or a service provided, the customer is the most common Failure Identifier. Customers can be broken down into several subsets:

1. General public
   a. Purchasers of high-price items or services
   b. Purchasers of low-price items or services
2. Government customer

Negative customer feedback could quite easily cause an employee to initiate the Financially Focused Quality process. Chapter 3 presents a detailed discussion on general customer feedback.

Government Failure Identification

Government customers differ greatly from the general public. Whereas the manufacturer of inexpensive headphone radios (discussed in Chapter 3) may eventually get feedback on 1 of 100 failures in the field, most government contractors receive an amazing 100% notification. This 100% really is not so amazing when you realize that many of the individuals (for example, military) inspecting for discrepancies have but one reason for employment by the government. They are being paid a salary (and often provided room and board) for finding flaws. This is different from the manner in which feedback is gained for products sold to the general public. The public is not paid directly for inspecting the products they purchase. It is, however, to their benefit to have certain problems corrected. Only with government contractors such as the military do you have the ultimate form of inspection by the end-user.

Warranty Service Department

Employees in the Warranty Service department routinely come in contact with failures. These employees are usually employed by the retailer, distributor, or manufacturer. Failures can be revealed to them by way of telephone calls, over-the-counter returns, or mail. Often their perception of the failure may be quite different from that of the customer. It is not unusual to find that the stated problem is actually only a symptom of a more serious root problem with much wider implications.

Failure Identifiers in the Factory

Inspectors

Obvious Failure Identifiers include those whose title is "inspector": individuals employed in a manufacturing concern whose job instructions specify actions to ensure consistent quality.

Manufacturing Employees

Although it is not generally in their charter, manufacturing employees in the shop are in an ideal situation to identify areas for potential improvement. It might very well have been a manufacturing employee who, while sitting around watching an inspector, thought to himself, "Hey, I can do that!" With that thought transferred to an FFQ document (Process Improvement Recommendation), he becomes an Opportunity Identifier, and tremendous savings can result (see Chapter 5).

All Other Company Employees

Many companies and corporations have internal audit organizations whose functions include formally investigating different areas of the company to determine areas where improvement is needed. Yet, with all employees acting as company auditors while performing their routine tasks (as FFQ suggests), the company should experience a much better audit, encompassing the entire corporation everyday.

Summary

Any human being can be a Failure or Opportunity Identifier. So, too, can a machine or computer designed and/or programmed to identify anomalies (defined as areas for potential improvement). In the Financially Focused Quality context, the goal is to:

1. Utilize every conceivable source for failure and opportunity identification.
2. Enter this data into a Financially Focused Process Improvement activity.

Financially Focused Quality is initiated via processing the appropriate form: Failure Notice or Process Improvement Recommendation

Failure Notice

The Failure Notice (see Figure 10.3) is used to initiate a Process Improvement related to delivery of specific products or services. The Failure Notice is used to formally cite the existence of a problem or defect. A form similar to that shown in Figure 10.3 could be modified and used for virtually any failure.

Failure During Manufacturing

For a failure occurring during the manufacturing process, the FN contains blocks for the following key elements:

1. FN form number
2. Date of failure
3. Failure Identifier information, including the name, organization, and telephone extension
4. Failed hardware information, including the nomenclature of the failed hardware and serial number and lot number, if applicable

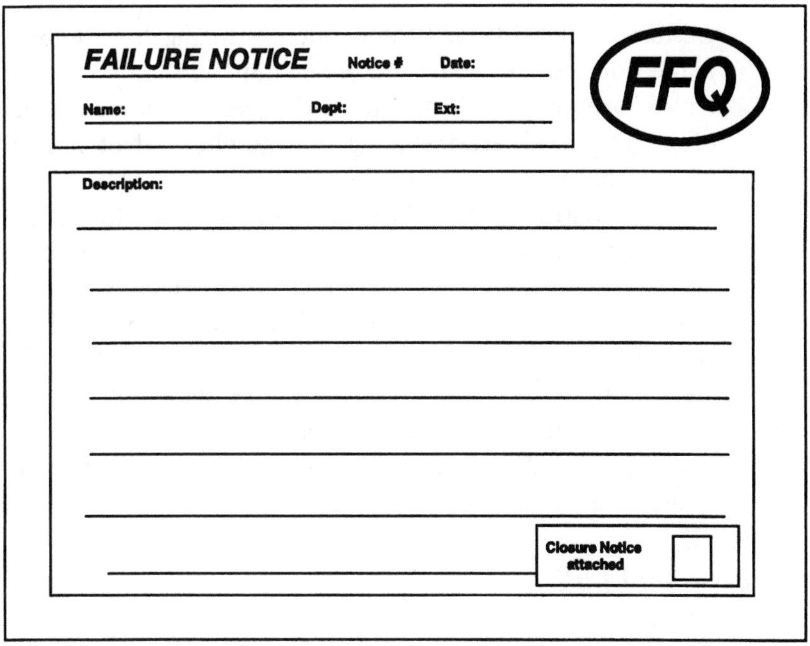
FAILURE NOTICE Notice # Date:
Name: Dept: Ext:
FFQ
Description:
Closure Notice attached

Figure 10.3. Failure Notice

5. Narrative regarding the failure, such as measured specifications, required specifications, conditions of test/inspection, unusual circumstances, and any other information felt to be relevant (for example, the Failure Identifier also is encouraged to list preliminary potential failure analysts)
6. The Process Improvement Closure Notice block, to be checked off when the Closure Notice has been issued

Failure: Customer Return

For the failure of a manufactured item returned to the manufacturer by a customer, the form includes the following data:

1. FN form number
2. Date of failure
3. Company Failure Identifier information, including the name, organization, and telephone extension

4. Customer Failure Identifier information, including the customer's name, address, area code, and phone number
5. Failed hardware (product) information, such as nomenclature of the failed hardware, including serial number and lot number of failed hardware, if applicable.
6. Narrative regarding the failure, such as what the customer says happened, conditions of customer use, the results of any tests performed by the company Failure Identifier (customer service representative) upon receipt of the defective unit, unusual circumstances, and other information felt to be relevant
7. Steps taken by the customer service representative (dependent on the warranty policy of the company)
8. The Corrective Action Closure Notice block, to be checked off as required

Failure: Hotel Industry

For failures related to the hotel industry, the FN contains blocks for the following information:

1. FN form number
2. Date of failure
3. Information about the Failure Identifier, which most likely will be a maid, desk clerk, PBX operator, cashier, or a reservation agent, as well as inclusion of the name, organization, and supervisor.
4. Failure information, such as location of failure (e.g., room number) and a narrative regarding the failure, such as (a) television is not functioning properly, (b) wake-up call buzzer will not shut off, (c) room heater/air conditioning will not operate, (d) geckos (lizards) are running around in guest rooms, (e) any other problems noted by guests, and (f) other information felt to be relevant. For example, the Failure Identifier is also encouraged to list preliminary potential failure analysts.
5. The Corrective Action Closure Notice block, to be checked off when the situation has been corrected.

Generic Failure

The generic Failure Notice can be described as having the following five key elements:

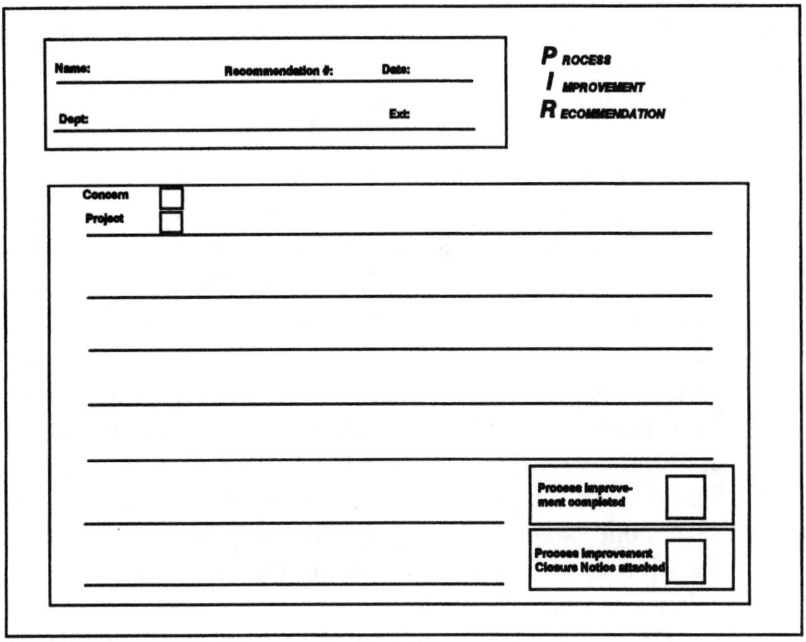

Figure 10.4. Process Improvement Recommendation

1. Failure notice number
2. Company identifier data
3. Customer identifier data
4. Nature of problem
5. A Process Improvement Closure Notice block, to check off when the failure has been corrected

The second form to be used for opportunities in Financially Focused Quality is the Process Improvement Recommendation.

Process Improvement Recommendation

The Process Improvement Recommendation (PIR) form is similar to forms utilized in many employee suggestion systems (see Figure 10.4). The recommendation form is a valuable tool for facilitating and motivating employee participation in Financially Focused Quality. A major advantage of the PIR is its ease and simplicity of use — FFQ makes it easy for every employee to participate in Process Improvement activity by offering the tools and techniques that are effective with minimal implementation cost.

The PIR will include the following:

1. Name of the employee appears on the top of the form.
2. A unique PIR number is preprinted on each form
3. The date of the submission is noted, which is important because, in the case of similar offerings, the PIR with the earliest date takes ownership of the recommendation. This does not mean, however, that those submitting similar ideas are not able to participate in the recommendation development process. Individuals with similar thoughts should be brought together, if appropriate, for the synergistic exchange of ideas.
4. The organization of the submitter appears along with the name.
5. The phone extension also appears on the form.
6. Two blocks are used for classifying the recommendation as either a concern or a project:
   a. *Concern:* When an employee observes a situation for which there could be potential improvement, the "Concern" block is checked, and the idea is written in a simple paragraph or two. There is no further work required on his or her part. That's it. It couldn't get much simpler, could it? Yet, by performing this simple task, management may learn of some key information, which could save millions of dollars for the company.
   b. *Project:* Occasionally, an employee is fortunate enough to have the ability to implement his or her own recommendation. When this is the case, the employee can undertake the project, and document related efforts with a PIR. Similar to the concern process, the block labeled "Project" is checked. The submitter proceeds to write one or two paragraphs about the project being undertaken, explaining what will be done and how the project will benefit the company. Again, what could be easier?
7. Finally, the PIR contains two final blocks on the bottom, one of which will be marked at a later date:
   a. *Process Improvement Closure Notice Attached:* As discussed later in this chapter, the Closure Notice formalizes procedural changes and closes activity associated with the PIR.

b. *Process Improvement Completed:* This block is marked with
    a check mark when the recommendation discussed on the
    PIR has been completed, and no further explanation is
    required.

Process Improvement Recommendation Success Factors

The three key reasons that the PIR is such a successful FFQ tool are
discussed below.

Ease of Processing

Unlike many suggestion systems or cost-savings programs, there is no
requirement to perform extensive analysis or calculations before the
recommendation is submitted. It takes only a few minutes to complete
the form. Also, because the form is submitted to one's own manager, the
following benefits should be realized:

1. The person receiving the form will be familiar with the termi-
    nology used by the submitter.
2. The person receiving the form will not reject it because it was
    not filled out perfectly. Many coordinators of traditional pro-
    grams take pride in having forms filled out properly, and much
    time is wasted in administrative detail. A manager is motivated
    to use time effectively. Someone who already has a full-time job
    managing the organization most likely will not be wasting time
    on meaningless details.

The ease with which the required FFQ paperwork is processed is a true
motivator for participation.

Low Implementation/Administration Costs

The group responsible for such programs in many companies often is
comprised of full-time employees, with designated part-time representatives in organizations throughout the company. These individuals
meet periodically (e.g., monthly) to review activities during the previous
time period.

Such meetings may turn into social affairs, with the serving of coffee
and doughnuts, while attendees determine, among other things, which

participating employees should be recognized for outstanding achievements. And such recognition, as mentioned earlier, is usually through the awarding of prizes and money. Tremendous amounts of corporate budgets fund this type of program.

Let's now examine the costs involved in administering the PIR. Each manager (whose time can be quite expensive) must read a one- or two-paragraph statement from an employee. This is time well spent for the following reasons:

1. The employee achieves visibility with management — a motivator.
2. The manager gets to know the employee better, which can lead to better utilization of manpower.
3. And, to top it off, the employee actually may have a beneficial idea, which could lead to significant savings for the company.

The mandatory management involvement is the only significant cost, and this involvement reaps rewards not found in traditional programs. More than offsetting the cost of management time is the elimination of the following customary costs:

1. Full-time program administrators and staff for administration of the previous program.
2. Part-time participation of organizational program support
3. Associated nonlabor costs, including:
   a. Publication and presentation costs
   b. Refreshments at meeting/luncheons/etc.
   c. Prizes, awards, and cash
   d. Travel expenses related to training for program staff

The bottom line is that the Financially Focused Quality program outlined above is much less costly to administer than other programs.

Mandatory Management Involvement

The traditional quality and productivity improvement programs (e.g., Total Quality Management, continuous improvement) declare that management support is *necessary*; however, FFQ goes that crucial step further by expounding the principle that management participation is *mandatory* by procedurally requiring that the manager interact with involved employees. This is another way FFQ cuts costs but increases effectiveness in suggestion programs.

It is not practical to involve an individual actually assigned to some type of Finance organization for every conceivable Corrective Action and Process Improvement. For those Corrective Actions with potentially wide-ranging financial implications, FFQ undeniably requires that a well-versed representative of Finance be involved. However, for recommendations involving low cost levels, a manager who has received the fundamental FFQ financial training should be able to represent financial interests and determine cost effectiveness.

As discussed earlier, most corporate suggestion systems and cost improvement programs require costly administration, often involving individuals dedicated to administering the program. In addition, these programs have large budgets for promoting such concepts as continuous improvement and for offering significant awards (e.g., prizes, money) to those participating in the program.

An important drawback of the above-mentioned legacy programs is that ideas are initially submitted to the individuals administering the program In most cases, management only gets involved in these traditional programs after the fact, after the improvement has been achieved. When the contribution is approved, the employee is recognized openly among peers in a management-conducted ceremony.

The individuals running the program are not as motivated to assist the employee submitting the suggestion as that employee's manager would be. Traditional programs require that detailed forms be submitted to higher authorities, which is a major problem because the individuals are not, in reality, higher authorities. One's manager truly fits this definition.

Financially Focused Quality procedures require that PIRs be submitted to the employee's manager. It is the responsibility of the employee's manager to pass judgment or coordinate review of the recommendation. When the recommendation has been written as a concern, the manager will log it into a master file. The next step is to make contact with management in the affected areas and discuss the potential of the recommendation. If a good probability exists for an improvement, the manager can seek additional information from his or her employee and organize appropriate contacts to continue the investigation.

For PIR projects, the manager will log the recommendation into the master file and discuss the advantages and disadvantages of the project with the employee. If the idea is deemed worthwhile, the employee begins the activity, and management periodically checks on progress. It may be concluded that perhaps the project is not cost effective. In this case, two minutes are spent completing a Process Improvement Closure Notice, the "Process Improvement Closure Notice Attached" block on the PIR is checked, and the Closure Notice is filed with the PIR.

Last, but by all means most important, the manager is directly responsible for the primary benefits an individual receives from employment. The manager offers promotions, raises, bonuses, training opportunities, and other benefits that traditional quality programs do not. The manager should include FFQ participation in the employee review process (see Process Improvement Recommendation Motivation section below). With the final ingredient of mandatory management involvement, the recommendation becomes a vital tool for most business entities.

Process Improvement Recommendation Motivation

A.H. Maslow theorized that employee motivation results from five broad classes of needs. These needs are arranged in a hierarchicy so that when one level is satisfied, the next level is activated. The five levels are

1. Physiological needs
2. Security or safety needs
3. Social needs, belonging, or membership needs
4. Esteem needs, including esteem of others and self-esteem
5. Self-actualization or self-fulfillment needs

The fact that the manager is reviewing the corrective action request provides a means by which all five levels can potentially be satisfied. Employees feel that they have much more control over their environments when they can easily submit suggested changes to management. The synergistic paradox suggests that the more control an employee has over his work environment, the more self-esteem he will hold.

Of course, Maslow's final need, that of self-actualization or self-fulfillment, depends on many factors that may be out of management's control. Yet, effective management has opportunities to influence the lives of their employees positively.

Another method of motivation worth mentioning here is almost in contradiction to one of Deming's 14 points listed in his theory of management (see Chapter 2). Deming states that companies should eliminate management by numbers and numerical goals. Yet, FFQ offers a reminder that there can be positive benefits associated with setting numerical goals.

Take the pole-vaulter, for example. The pole-vaulter loves his sport. It is almost impossible to express the power felt by the pole-vaulter as he stands on the runway, grasping his pole tightly and eyeing that bar high up in the sky. A deep breath is taken and then slowly exhaled. The pole-vaulter looks down with determination and begins the first steps down the path to the pit. He is picking up speed, going faster and faster. It is poetry in motion as the pole is planted, and the vaulter gracefully rises up in the air, sailing like a picturesque seagull. Up, up, and finally over the bar, landing — exhausted — on the foam-rubber padding in the pit. The pole-vaulter's eyes are closed. He felt his foot touch the bar as he went over. Did it fall? He's afraid to open his eyes. He hears the fans in the stands cheering loudly. He opens his eyes, and *the bar is still there!* He did it. The jubilant pole-vaulter stands and jabs his fist triumphantly into the air.

Nice story, yes. But, wouldn't you agree that a significant amount of the story's impact would have been lost if, prior to the vault, a track meet official had removed the bar? Don't you feel, too, that the pole-vaulter would lose much of his motivation without the bar — without a goal? The financial focus encourages management to offer goals to employees — goals tied to PIR submittal. Such goals can be established for individuals or entire groups of employees as follows:

- *Groups of employees:* A work group would strive to submit a certain number of valid recommendations by a certain date or to submit more recommendations than another group. The prize for achieving the goal number or for submitting more than the other group could be a group picnic. Such an approach would provide such benefits as encouraging teamwork and fostering a healthy competition between working groups.
- *Individual employees:* Inform employees that the number of PIR submittals will be a factor in the performance review process, which could have both positive and negative effects. On the positive side, employees who normally would not participate in such an activity will feel more compelled so do so; however, there may be some employees that feel pressured and object, saying: "It isn't my job!" For such employees, management is

encouraged to offer an explanation along the lines of: "You are right. It is not your job to cut costs and work more efficiently. In fact, it is my job. But your job is to help me do my job, and I'd really appreciate your trying your best to improve our company's cost effectiveness. You can show me that you are trying to help by looking for and submitting some recommendations. You can do it!"

Examples of Process Improvement Recommendations

Recommendations can have extremely beneficial impacts on a business concern. Below is a sampling of simple recommendations that resulted in significant savings. Each recommendation first appeared in the one- or two-paragraph narrative form.

1. Add "was" and "is" columns to existing report to ease identification of changes.
2. Take advantage of state-of-the-art technology to improve communications.
3. Centralize financial control of low-value activities.
4. Utilize a 12-month moving average for forward pricing rates, instead of formal forward pricing rate negotiations.
5. Generate contract-unique rates for estimates to complete.
6. Budget and track common minor material expenditures.
7. Publish annual supplements instead of annually republishing an entire text.
8. Automate course presentation to eliminate requirement of having a teaching assistant.
9. Consolidate impact studies into one large report instead of preparing and distributing 35 small reports.
10. Discontinue maintenance of unnecessary documents.
11. Discontinue performance of unnecessary tasks.
12. Generate equivalent rates for estimates to complete.
13. Create a model 204 computer program for generation of reports previously performed manually.
14. Utilize numerous word processing applications.
15. Computerize exception report analysis.
16. Discontinue preparing and including unnecessary data in certain reports.

17. Consolidate variance reporting.
18. Include rate-volume analyses on existing reports.
19. Directly transfer data files from one system to another.
20. Utilize badge entry units for access to buildings with special security requirements.
21. Take advantage of bar-code technology for inventory functions.

Illustration of Acting on a Process Improvement Recommendation

To illustrate the ease of processing a PIR, the following case is presented. In this organization, the manager has explained to employees the PIR process and has requested that each employee submit at least one PIR each quarter. One employee who is responsible for performing a series of monthly reports learns of an accounting change which will cause her to make several procedural changes.

Traditionally, the employee would make the changes in report formats and procedures as time permitted. Because in most business environments there is a tendency to reject change, the employee may avoid taking corrective action until the last minute and even then implement such changes reluctantly. In this case, however, the employee learns of the coming accounting change, and ...

1. She takes a PIR form from the desk.
2. She fills out the heading information.
3. She puts a check mark in the "project" block.
4. She writes the following: "A consolidation of overhead pools will be effective month-end August. I will no longer need to report two sets of data; therefore, the following steps should be taken: (a) modify the Overhead Report accordingly, and (b) modify accordingly my job instructions and procedures."
5. The completed PIR is copied, with the original going to the organization manager. The employee retains the copy for reference.
6. The manager receives the PIR and (a) reviews the PIR, (b) interacts with the submitter, (c) coordinates meetings and research towards the goal of improving company cost-effectiveness, (d) ensures that the PIR is eventually closed either with a Process Improvement Closure Notice or with a check mark in the "Corrective Action Completed" block, and (e) utilizes records

regarding employee FFQ involvement in the periodic performance appraisal process.

In the above procedure, the following benefits are easily realized:

1. The employee creates an action item, thus has defined a goal.
2. The manager's awareness of the employee's activities provides management with the opportunity to better evaluate employee performance.
3. As a result of receiving management attention, the employee has an incentive to perform the task in a thoroughly competent fashion.

The task most likely will be given a high priority by the employee and will be accomplished in a timely manner. Note that, in this model, the savings are not easily quantified. This situation is quite acceptable within the realm of FFQ. Any time Corrective Action or Process Improvement costs are less than the benefits derived, the Process Improvement activity has been successful.

Summary

This chapter has discussed the many methods for identifying failures and ensuring that the other areas for potential improvement are targeted, as well. Once the appropriate FFQ documentation has been generated, the cycle continues with the coordination and eventual selection of a cost-effective Corrective Action (see Chapter 11).

Self-Study/Discussion Questions

1. If your boss encouraged you to make recommendations, how would you feel? Do you think he would take your recommendations seriously? Or would you think: "Well, here's another one of those suggestion programs"? What would it take to really motivate you to participate?
2. Have you previously participated in a suggestion program? Share your experience.
3. For what business process improvements have you personally been responsible? How were you rewarded for your efforts?

11 Financially Focused Quality Process Improvement Coordination and Decision-Making

Introduction

The previous chapter presented a detailed description of how and where opportunities for process improvement are identified. It also provided recommendations for enhancing visibility in all areas where improvements might be achieved. The next step is financially focused coordination and resolution of the opportunities identified.

Process Improvement Coordination

The Process Improvement Coordinator is that individual responsible for ensuring that, once an opportunity has been identified, or a failure has been determined, the financial focus is applied to the process. Figure 11.1 presents the following Process Improvement Coordinator functions:

1. Receives Process Improvement Recommendation (PIR) or failed hardware and Failure Notice (FN) from the Failure or Opportunity Identifier.

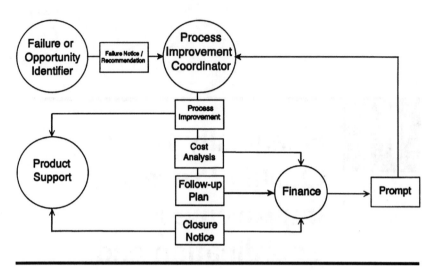

Figure 11.1. Process Improvement Coordination

2. Via many methods, generates the PIR and forwards it to the product support organizations taking action.
3. Generates the Process Improvement Cost Analysis (PICA) and forwards it to Finance.
4. Generates the Process Improvement Follow-Up Plan (PIFP) and forwards it to Finance.
5. Generates the Process Improvement Closure Notice (PICN) and forwards it to the Finance and product support organizations.
6. Receives prompts regarding PIR effectiveness, cost estimates, and follow-up plan from Finance.

When Process Improvement or Corrective Action Is Not Required

Many times a reported failure or nonconformance would not require a Corrective Action. A sample of such a situation would be when nonconformance is documented to meet procedural requirements (e.g., material shelf-life expired but analysis reveals that the product still complies with specifications), and no Corrective Action is needed. A shelf-life extension should require a discrepancy record and test prior to an "extend shelf-life" disposition. Also, no Corrective Action is required when unavoidable damage or defect is caused directly by the performance of specific rework or repair activity, and the final product meets desired standards.

Time Requirements for Process Improvement Coordination

Many corrective action and productivity improvement programs in large companies utilize a full-time individual or team to administer the program or system. With FFQ, only one individual has the responsibility of coordinating activities related to a particular recommendation or failure. This position, appropriately known as the Process Improvement Coordinator, is never occupied on a full-time basis. In fact, the primary goal for this individual is to facilitate the most cost-effective Corrective Action in the most efficient manner possible.

The Process Improvement Coordinator should be the manager of the employee submitting the recommendation; however, when there is a product failure, Process Improvement coordination responsibility is often delegated to employees in the Quality Engineering organization. Toward the goal of achieving top quality at a minimum price, the coordinator makes use of proven FFQ tools and techniques.

Time Requirements for Process Improvement Recommendation Coordination

As discussed in Chapter 10 (Process Improvement Recommendation Success Factors), the manager of the employee submitting the recommendation is responsible for its coordination. The requirement of coordinating individual PIRs on a part-time basis should not be a significant problem because management can analyze the workload of employees and delegate this responsibility. However, the manager must maintain the log and be actively involved in the closure of each PIR to ensure that the benefits of management involvement listed in Chapter 8 are realized.

Time Requirements for Failure Notice Coordination

The manager is still the key player for major Failure Notices. Depending on the environment (e.g., aerospace and defense manufacturing), literally hundreds of FNs can be generated every week. As a result, it is not practical for management to be involved in every minor situation.

Whenever possible, coordination of Corrective Action and Process Improvement action should be performed by the individual causing the failure. Minor FNs for scratches, dings, loose moldings, etc. should first be brought to the attention of the supervisor of the manufacturing area performing the related work.

After a failure mode has been determined or if the cause of the failure cannot be identified, responsibility for coordination is assigned to the appropriate manager. Even in these situations, the manager may delegate coordination activity to another. Still, management is required to be involved in the FN closure.

Process Improvement Recommendations: Related Functions of the Process Improvement Coordinator

As discussed in the previous section, the Process Improvement Coordinator, when coordinating activities related to recommendations, will always be the manager of the employee making the submittal. Again, in this situation, the manager performs the following functions:

1. Reviews the recommendation.
2. Interacts with the one submitting the recommendation.
3. Coordinates meetings and research towards the goal of improving company cost-effectiveness.
4. Ensures that the recommendation is eventually closed either with a PICN or with a check mark in the "Process Improvement Completed" block.

With the exception of reviewing and eventually closing the PIR, all of the above activities should relate directly to existing management functions. Routine managerial functions overlap with the employee interaction process. Coordinating meetings and performing or authorizing research regarding potential improvements in policies and procedures should be an ongoing management exercise.

Financially Focused Quality, therefore, provides a framework for management to become directly involved in the Process Improvement and Corrective Action cycle. Such involvement becomes mandatory where the PIR is applied.

The Financially Focused Quality Mindset

A key to successful Process Improvement activity is acquiring and maintaining the Financially Focused Quality mindset. This mindset requires that all significant financial implications be considered before taking action to correct a failure or improve upon an existing process, policy, procedure, or situation. Remember that in a company where Financially Focused Quality has been properly implemented all employees receive financially focused training. This training increases the probability that the recommendations submitted to the PIC result in improved profitability.

It is not always practical for the coordinator to involve a member of the corporate financial community in the review of recommendations. Therefore, the next best individual is involved — the coordinator himself, who is a manager. The manager, by the nature of the job description, should spend a good deal of time maintaining a consciousness of the financial aspects of his organization. Perhaps the biggest factor contributing to this concern is the fact that most managers negotiate or are given annual budgets with which to operate their organizations. In this regard, all company management is already prone to the Financially Focused Quality mindset. This is a key reason why members of management coordinate recommendation processing.

The Process Improvement Coordinator and the Failure Notice

The person assigned to be the Process Improvement Coordinator responsible for processing the Failure Notice will vary depending upon the nature of the failure. Below is a sample listing of failures, Failure Identifiers, and probable Process Improvement Coordinators for various environments.

Auto or Aircraft Manufacturing Failure

Cracks in manifold flanges are observed during the final assembly process. An employee in the final assembly area spots the failure. The supervisor of the final assembly functions is notified of the failure and is designated as the Failure Identifier. Upon learning of the cracked manifold flange, the supervisor completes the Failure Notice. While processing the notice, the identifier considers the likely cause of the failure. If it is felt that the failure has resulted from an error in the final assembly process, he may very well designate himself as the Process Improvement Coordinator and take responsibility for closure of the Failure Notice.

In this case however, the Failure Identifier feels that this problem has the potential for having major impacts on the product. Similar failures in the past lead him to believe that this failure mode is related either to the materials or to the processes used for manufacturing the manifold flanges. Therefore, Financially Focused Quality begins as the notice is forwarded to the manager who is responsible for flange manufacturing. The Process Improvement Coordinator is the manager responsible for flange manufacturing.

Television Manufacturing

Television sets are being returned because the volume controls do not work. The consumer Failure Identifier is the individual who recently purchased the television. The company Failure Identifier is an employee in the Warranty Service department. That employee, upon receiving a returned defective television, completes the Failure Notice and forwards it to the Process Improvement Coordinator. The Process Improvement Coordinator is the manufacturing engineer manager.

Service Industry Failure #1

The faucet in the bathroom of Suite 568 is leaking with an annoying, steady drip. The customer Failure Identifier is the guest who just checked in to that room. The company Failure Identifier is the front desk clerk answering the phone when the hotel guest calls. The clerk completes the Failure Notice and forwards it to the Process Improvement Coordinator. The Process Improvement Coordinator is the engineering manager.

Service Industry Failure #2

There is an ugly stain on the carpet in Suite 568. There is no consumer Failure Identifier. The company Failure Identifier is the maid who cleaned the room after the sloppy guests checked out. The maid completes the Failure Notice and forwards it to the Process Improvement Coordinator. The Process Improvement Coordinator is the manager of the housekeeping department.

The Coordination Process

Ideally, the coordinator would only have a few PIRs or FNs to deal with at any one time, but there will be situations where there are many related

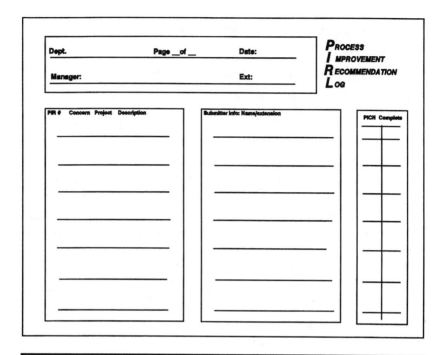

Figure 11.2. Process Improvement Recommendation Log

failures. There may also be many recommendations submitted to the same coordinator. As such, the coordination process begins with logging in the FN or the PIR on a simple log-in form. An example of a tracking form used for recommendations appears in Figure 11.2. This form is affectionately called the PIRL (Process Improvement Recommendation Log). The PIRL easily provides a summary of the recommendation activities that are ongoing in an organization. The PIRL doubles as a readily available action item list to which management may refer when there is some free time.

Coordination of Process Improvement Recommendation Projects

In the case of PIR projects, the manager uses the log to interact with and check on the progress of the pertinent employees. This tool increases the employees' awareness of management's concern for their work. In turn, the management activity fosters motivation and pride in performance.

Coordination of Process Improvement Recommendation Concerns

Here is the procedure used for PIR concerns. If the concern is of such a nature that the manager can directly implement it, then:

1. The manager will analyze the pros and cons of the recommended action. Very important here is management's assessment of cost involved in the "before" and "after" environments. A valuable tool in performing this analysis is the Process Improvement Cost Analysis (PICA), discussed in detail later in this chapter. In the case of a concern (as opposed to a project), the decision regarding a recommendation's cost-effectiveness might be obvious to management trained in FFQ. In this case, there is no need to perform the PICA. The true value of the PICA, as discussed below, is for PIRs with relatively high cost impacts.

2. If management deems it necessary, appropriate research and analysis will be conducted.

3. A final decision will be made in a timely manner.

4. If the recommendation is adopted:
   a. Necessary steps to ensure its implementation (e.g., policies and procedures updated, responsibilities assigned) will be outlined on the Process Improvement Closure Notice (discussed later in this chapter), and those activities completed.
   b. When the Closure Notice steps have been completed, the notice is attached to the PIR, and the "PIR Closure Notice Attached" box on the recommendation form is checked.
   c. Management informs employee of recommendation implementation. Depending on the magnitude of resulting benefits to the organization, the manager may reward the employee (e.g., commendation, bonus).

5. If the recommendation is not adopted:
   a. The decision is discussed with the submitting employee.
   b. Rejection justification is written on the original PIR (if necessary, another page is attached), and the "PIR Completed" box is checked.

If the concern is of a nature that must be coordinated with other organizations, the manager:

1. Chooses process analysts.
2. Prepares background information and sends it to the analysts for their comments.

Upon receiving input from the analysts, the manager may feel that adequate data exist to evaluate the recommendation. The following steps are performed if the manager feels more data are needed:

1. The manager invites the analysts and/or a representative of the Finance organization to the process analysis meeting.
2. The manager facilitates the meeting.
3. If appropriate, the manager coordinates the use of the Process Improvement Cost Analysis (PICA).
4. The manager summarizes the meeting.
5. The manager ensures a final decision is made in a timely manner.
6. If the recommendation is adopted:
   a. Necessary steps to ensure its implementation (e.g., policies and procedures updated, responsibilities assigned) will be outlined on the Corrective Action Closure Notice, along with those activities completed.
   b. When the Closure Notice steps have been completed, the notice is attached to the recommendation form, and the "Closure Notice Attached" box is checked.
   c. Management informs the employee of the recommendation's implementation. Depending on the magnitude of resulting benefits to the organization, the manager may reward the employee (e.g., commendation, bonus).
7. If the recommendation is not adopted:
   a. The decision is discussed with the employee who submitted it.
   b. Rejection justification is typed onto the original recommendation form (if necessary, another page is attached), and the "PIR Completed" box is checked.

If the concern is of a nature over which the manager has no authority:

1. The manager forwards the recommendation to the manager with the power to implement the change.
2. That manager follows the procedures above and comes to a timely disposition, closing with both the submitting employee and manager.

Coordination of Failure Notices

The Process Improvement Coordinator assigns investigation of the notice to an employee with appropriate expertise. Based on the outcome of this investigation, a process analysis meeting may be necessary, requiring the following steps:

1. Potential failure analysts are selected.
2. Background information is prepared and sent to the analysts.
3. The analysts and, if the coordinator does not feel adequately trained, a representative of the Finance organization are invited to a process analysis meeting.
4. The coordinator runs the meeting.
5. If appropriate, the Process Improvement Cost Analysis (PICA) is used.
6. Meeting decisions are summarized.
7. More meetings in smaller groups may be needed, but eventually the final PIR is determined.
8. Necessary steps to ensure Process Improvement implementation are outlined on the Process Improvement Closure Notice (PICN), and those activities completed.
9. A Process Improvement Follow-Up Plan is prepared.
10. When the Closure Notice steps have been completed, the "PIR Closure Notice Attached" box is checked, and the notice is attached to the original recommendation form.

Failure Analysts

In the quest to determine the individuals who can really contribute to resolving failures in a cost-effective manner, the Corrective Action Coordinator compiles a list of potential failure analysts which could include all individuals having even the slightest probability of contributing to failure resolution, such as individuals involved in the manufacturing process or employees working in the packing, shipping, and/or shipping/receiving areas. Subcontractors providing materials for use in manufacturing should also be considered as potential failure analysts.

The coordinator will send a copy of the FN to all potential failure analysts that may have legitimate involvement with the failure reported on it. Upon receipt of the notice, the analysts review the summary of the failure and then try to determine any way in which their function may have contributed to the failure. After a thorough internal analysis of their functions, they next address other potential causes, such as when they may have detected an error caused by another organization in the production process. Providing this feedback need not be time consuming, and often a phone call or brief note suffices. The key here is to eliminate unnecessary effort. Sometimes, too, this process will result in very timely correction of the failure, avoiding lengthy sessions of misusing time with superfluous discussion. Financially Focused Quality recognizes the waste that can result from taking time to educate and then include uninvolved parties in discussions.

Each potential failure analyst provides feedback to the Process Improvement Coordinator, and perhaps even more failure analysts are identified. The inputs from analysts and newly identified analysts are carefully reviewed, and the coordinator narrows down the list of analysts to assign a final team. In order to ensure a cost-effective Process Improvement cycle, only those individuals who truly are capable of positively contributing to the analysis should be involved in time-consuming investigations. To summarize:

1. By starting with a listing of all potential failure analysts, the total picture can be studied to ensure that nothing is missed.
2. Sometimes a Process Improvement will be obvious to one of the analysts, and the time involved in developing a Corrective Action can be substantially reduced.
3. Potential failure analysts with a low probability of contributing to the Corrective Action may not need to be involved.
4. Many in industry are very much aware of the unnecessary meeting syndrome (i.e., many people being asked to attend meetings from which no benefits are received). With the FFQ procedure, only those individuals who are needed or have a reasonable probability of being able to help with failure resolution are involved.

Once the final listing of failure rectifiers is assembled, the Process Improvement Coordinator has several options. For starters, the coordinator can meet individually with Failure Identifiers. A second choice is to meet with Failure Identifiers in small groups.

For major failures with far-reaching implications, a process analysis meeting may be appropriate. When such a meeting is deemed essential, the Process Improvement Coordinator compiles essential background information that is forwarded to analysts so that they may prepare before attending the meeting.

Process Analysis Meeting

The Process Improvement Coordinator ensures that the final failure analysts obtain copies of the FN and are notified as to the time and location of the meeting. The paperwork accompanying the notice includes a worksheet on which the analysts can list potential causes and Corrective Actions or Process Improvements to bring to the meeting or discussion. This meeting launches the process whereby a financially focused Process Improvement is generated, yielding the highest quality at a minimum cost. A sample meeting agenda follows.

Attendees are welcomed and introduced. (*Note:* attendance at the meeting is limited to only those individuals who genuinely should be there.) They might include:

1. Failure analysts, from the following areas:
   a. The Warranty Service department (correcting problems themselves) — often proper training of these employees can lead to them being able to correct failures before the customer leaves. For example, if a television set is being returned because the customer misunderstood instructions, a knowledgeable customer service clerk can eliminate the need to begin the Process Improvement cycle. Of course, such instructions may need to be rewritten to preclude future misunderstandings.
   b. Production/Manufacturing
   c. Production support (e.g., expediters/dispatchers)
   d. Suppliers (e.g., of materials and subassemblies)
   e. Quality Engineering
   f. Customers
   g. Training
   h. Management
2. Financial representatives:
   a. Industrial Accounting
   b. Cost Accounting
   c. Program Controls

3.   Other employees:
  a.   Process Improvement Coordinator
  b.   Appropriate levels of management
  c.   Failure Identifiers
  d.   Any others who may be necessary

Again, it must be emphasized that, for ultimate cost-effectiveness, only those individuals with probable need to attend the meetings should be invited. For example, someone may identify the failure during shipping inspection. It may be determined that the attendance of this inspector is not necessary for the Corrective Action process.

In the case of a failure identified during the manufacturing process, however, it may be believed that the Failure Identifier is actually part of the failure's cause. If this is the case, this individual is critical to the Corrective Action process.

The remainder of the meeting will consist of the following:

1.   The background of the failure is reviewed, while allowing questions from the attendees to ensure their complete understanding of the situations.
2.   Repercussions of the failure are discussed (e.g., potential impacts to other products, etc.).
3.   The objectives of the meeting are discussed. The obvious purpose here is to discover the cause of failure and/or identify a failure mode, but, at the same time, other goals could include improving quality, safety, and communications; cost reductions; and improvement of the company's competitive position and employee job security.
4.   To ensure an efficient meeting, the coordinator should provide a brief statement about what is not relevant to the meeting.
5.   Although some situations may require discussion of the following issues, in most cases they should be avoided:
  a.   Pay rates (labor rates/salaries)
  b.   Grievances (administration of overtime, etc.)
  c.   Personnel matters ("Why doesn't Charlie have to empty wastebaskets?")
  If the above issues do not play a part in the Corrective Action identification process, it should be pointed out that there are other channels for dealing with such matters.

6. The coordinator asks each analyst, one at a time, to report on their study of the failure according to the analysis conducted upon receipt of the Failure Notice background data.

7. The coordinator records each pertinent point on a chart (Vu-Foil, whiteboard, etc.) in front of the room.

8. During this initial presentation of each analyst's statement, the only comments permitted from attendees are questions intended to clarify. Attendees should not be defensive or judgmental during this part of the meeting.

9. After the analysts have made their presentations, the coordinator explains the rules for the brainstorming session about to begin, and then the process begins (see below).

10. During the brainstorming session, a list of *potential* causes is generated. The next step is determining *probable* causes. The coordinator begins at the top of the list, and, one by one, the group addresses the potential causes. Some of those listed may obviously be eliminated and a line drawn through them; however, the coordinator should maintain the listing of those that have been eliminated so that they may be addressed at a later date if all of the probable causes have been eliminated.

11. Responsibility for each probable cause is assigned to an analyst. Ideally, this analyst will already be in attendance, but there is a chance that, over the course of the meeting, the need for new analysts has been identified. The responsible analysts will conduct a thorough analysis and report back at the next process analysis meeting.

12. A date, time, and place for the next meeting are announced.

13. The coordinator thanks attendees for their attention and cooperation and adjourns the meeting.

14. With the details of the meeting still fresh, the coordinator creates meeting minutes by updating the agenda with the events of the meeting and e-mails the minutes to all attendees.

Brainstorming

D.L. DeWar (1982) has defined brainstorming as "using a group of people to stimulate the production of ideas." DeWar provides an excellent analysis of brainstorming in his text *The Quality Circle Guide to Participation Management.* His analysis has been enlarged upon below to facilitate the Financially Focused Quality mindset

The definition of brainstorming provided above is very appropriate for its applications to Financially Focused Quality. FFQ involves targeting a process for improvement. If the process involves more than one individual, it is almost always more effective to brainstorm with a group of process performers than to have a single individual attempt to generate ideas alone.

With Financially Focused Quality, the targeted process is that which resulted in identification of an opportunity or a failure. The brainstorming session will revolve around the following four goals:

1.   Generate potential causes of the failure.
2.   Once all potential causes have been generated, determine probable causes of the failure.
3.   Identify potential Process Improvements.
4.   Prioritize the Process Improvements.

While the traditional approach to brainstorming leads to the generation of ideas, this text zeros in on the Financially Focused Quality-related goals of revealing causes and devising Corrective Actions. The success of the brainstorming process results from unlocking the creative power of the group participants. This process begins with the identification of a topic or objective.

Topic or Objective Definition

Prior to any brainstorming session, it is essential that an objective be identified. In FFQ, the topics will spin off from the specific failure(s) being addressed — for example, "Reception on television is not consistent with design specifications." However, to ensure maximum returns from the investment of time expended by attendees, it is essential that such descriptions be as precise as possible.

The above example relating to television signal reception could be more effectively stated as: "Fine-tuning apparatus operates intermittently." With this definition, the discussion will focus on procedures and hardware relating to fine-tuning.

Recording Ideas

Someone needs to be given the responsibility of writing the ideas as they are given. This is preferably performed via some means whereby attendees can see what has been written (e.g., Vu-Foils, charts, whiteboard, etc.).

Presentation of Rules

The brainstorming process is performed at its best when certain rules or guidelines (discussed below) are followed. Prior to each session, the Process Improvement Coordinator should review the rules he has selected to use with attendees. Although the rules may vary from session to session, a sampling of such rules is provided below. These rules also provide an outline of how a typical brainstorming session might progress.

Take Turns

The process begins with each member, in rotation, being polled for potential causes of the failure (new ideas). This process continues until all ideas have been exhausted. It is important for each member to offer only one idea per turn regardless of how many he or she has in mind. This allows ideas to be expressed by less-assertive individuals and encourages their participation. Of course, as the session progresses, some individuals will be at a loss for a new idea. When this happens, the participant merely says "pass", "I'm still working on it", or "you got me, dude", and the spotlight rotates to the next member.

Place Maximum Effort on Idea Generation

An energetic effort should be applied toward generating a large quantity of ideas, thus fully utilizing the effectiveness of the team process.

Don't Throw Anyone into a State of Self-Conscious Distress

This is perhaps the most important rule. It is important to remember that no idea should be treated as silly or stupid. Few people, if any, want to be ridiculed, and such belittling or criticizing will surely curtail the creativity of team members.

Everyone Has To Participate

Some members may not be particularly assertive or may even be shy. For these, it may take courage to start participating. Their ideas should be welcomed and such individuals reassured. Enthusiastic support of their ideas — and the ideas of everyone — is essential.

Relax and Have a Sense of Humor

This rule must be stressed. Brainstorming should be fun. Wit and a keen sense of humor are critical elements for a successful brainstorming session. Good-natured laughter (not at a team member's expense) and informality should be encouraged to enhance the climate for innovative activity. However, it should be obvious that derisive laughter will have an unwelcome and dampening effect.

Exaggerate

These sessions are one of the few business situations in which exaggeration should be encouraged. Such behavior may add humor, and it adds a strong creative influence to discussions.

Condense Ideas and Be Concise

This rule is intended to deter individuals from talking too much. Being concise will result in more efficient sessions. The leader should, whenever possible, edit a lengthy idea into a clear and concise statement. However, he must also ensure that the originator agrees with the condensation. For example, a member may state: "I've noticed something unusual about the way folks are soldering tuners to the chassis. Each person is using a different approach!" The coordinator obviously should not take the time to write the entire sentence of probable cause and should choose to paraphrase, writing something like "inconsistent soldering techniques."

No Judgments Allowed

While attendees are giving their ideas, absolutely no evaluation of the suggested ideas should occur. This rule also applies to the leader and includes positive comments as well as the negative ones. For example, "Where did *you* come from?" or "Hey, that's great!" are both examples of no-nos. No comments of any type are allowed during the brainstorming process — only clarifications.

Completion of Brainstorming and Discussion

The brainstorming is complete when no one is able to think of another idea. The next step is to select the probable cause. At this point, reasonableness enters the equation. Now is the time to evaluate, and even ridicule, if (and only if) appropriate. Some of the ideas generated during the session will most likely be preposterous. This is the time to narrow the list down to a more manageable level.

Traditional brainstorming methods expedite this process by using a simple voting technique. Voting is successful because the members are the experts in their areas. Each idea is voted upon, and members can vote for as many ideas as they feel have value. The leader records the number of votes next to the idea. Only votes in support of an idea are taken. There may be no votes against an idea.

Financially Focused Quality and Probabilities

A financially focused technique suggests that probabilities be incorporated into the brainstorming process. Once the list of potential factors has been narrowed down to the probable causes, the listing can be further prioritized. Members vote to assign a numerical value for the probability of each being a major factor.

The coordinator explains that each attendee should consider the list of probable factors and, based upon his unique expertise, estimate the percent of probability that each factor contributed to the failure. The totals of these probabilities determined by each member should equal 100%.

If the members feel it would be beneficial, the meeting could be adjourned at this time and scheduled to resume at some point in the near future. The break between meetings is intended to allow participants time to evaluate each potential factor and determine the probability of it being, in fact, the cause or a part of the cause of failure. With this approach, the coordinator polls each member individually, listing their probability percents next to the corresponding idea. After members have supplied their percents, the coordinator must add the totals to ensure they equal approximately 100%.

Selection

When using the traditional brainstorming method, a circle is drawn around the idea that receives the most votes, and attendees decide how many of the top ideas merit further consideration. With the financially focused approach, composite probabilities are determined, and the factors are numbered in order from highest probability to the lowest.

Members are now able to concentrate on a few important items instead of being potentially overwhelmed by many. In traditional brainstorming, the remaining ideas are again voted on, with each member getting only one vote this time. Rankings are written beside each of the circled ideas. To maintain a financial focus, this is the time to utilize the Process Improvement Cost Analysis (PICA).

Of course, a member may call for a suspension of voting while he makes arguments either for or against a particular idea. It is important that all data available at the time be made known to the attendees to ensure that the voting is as accurate as possible. As a result, again, all members are enthusiastically encouraged to participate in the discussion.

Components of Effective Brainstorming

For traditional brainstorming, when the topic has not been predetermined by way of a failure, the leader may wish to distribute an agenda prior to the meeting. This gives members a chance to prepare for the upcoming brainstorming topic and perhaps have some ideas generated, adding to a successful beginning to the meeting.

Some leaders approach brainstorming by using a large sheet of paper attached to a wall in the front of the room. Such a device makes ideas viewable to all attendees, and it can become a permanent record of meeting events. Such a sheet could effectively be used at a later time for presentations to management.

Members should be encouraged to look at the big picture. With such an approach, a new range of thoughts will result. For example, first there were tiny airplanes. Eventually, visionaries and engineers brought about DC-10s and 747s. Similarly, little rafts led to the development of gigantic ocean liners.

It is helpful to combine the old and the new. Combinations of existing ideas and concepts have been known to lead to exciting innovations. For example, early ships were powered by combining the steam engine and the paddle. Emphasize imagination and creativity. It has been said that creative thinking is often hampered by right-brain thinking. Imagining an impossible concept, for example, will often free the mind of such restrictions. Imagine that the sun never sets or, when brainstorming for solutions, imagine that your company has absolutely the best technical engineering staff in the world.

Miniaturization is an approach for trying to make things smaller, faster, or more concise. When ideas begin to get scarce, suggesting miniaturization may help members generate ideas in a new direction.

Incubation often takes place after an initial brainstorming session. If attendees feel it is appropriate, the meeting may be adjourned so that they might "sleep on it", and new ideas may emerge.

Ideally, only one brainstorming session will be necessary. This assumes that all team members are fully prepared. However, as discussed above, one or even two follow-up sessions may be required to ensure that all necessary data have been used for Process Improvement generation. Once the probable Corrective Actions have been developed, FFQ next calls for utilization of PICA.

Process Improvement Cost Analysis

As discussed earlier, PICA (Figure 11.3) can be a valuable tool for determining which actions to select in cases of Process Improvement Recommendations or concerns and in determining Corrective Actions related to FNs. This form may be generated with spreadsheet software and used by anyone considering alternatives; however, because the research and computations required for accurate analyses can sometimes be quite time consuming, judicious use of PICA is encouraged.

All Process Improvements and recommended procedural changes have financial implications that can be measured either directly or indirectly. For example, a Corrective Action or Process Improvement that results in savings of 10 minutes per test in the inspection process has savings that can be quantified as follows:

1. Hourly direct labor rate of inspector = $18 per hour.
2. Time saved per test = 10 minutes.

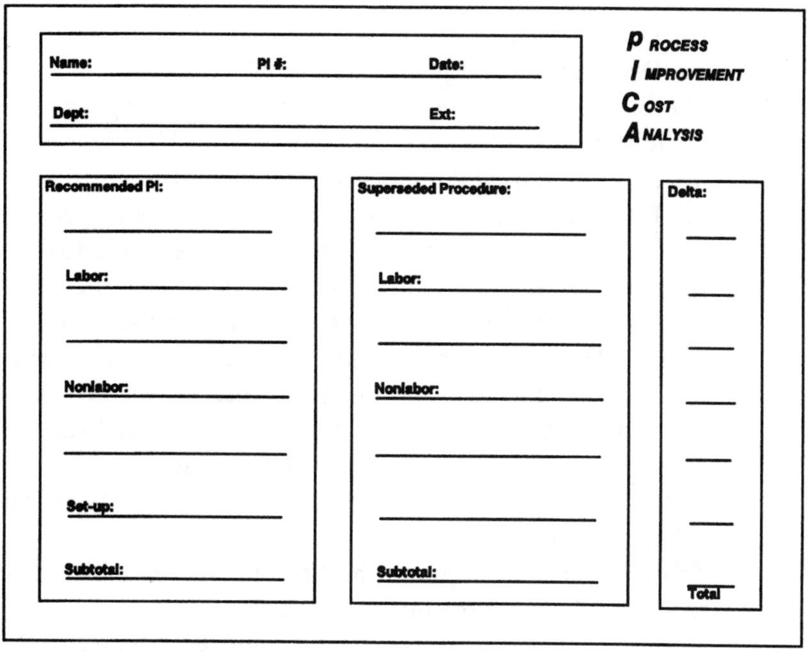

Figure 11.3. Process Improvement Cost Analysis (PICA)

3. The 10 minutes saved per test by the inspector = 1/6 of an hour.
4. 1/6 of an inspector's hour saved for each test = 1/6 of $18.
5. 1/6 of $18 (inspector's wage per hour) = $3 saved for each test.
6. Assuming there are 100 tests each month, the Process Improvement yields gross savings of $300 each month.
7. Costs associated with implementation should be subtracted from the gross savings, because certain one-time costs may be unavoidable (e.g., rewriting procedures or purchase of new equipment).
8. When appropriate, further computations may be performed to account for savings related to any elements of overhead involved in the cost.

A PICA is filled out for Process Improvement #1 as described below:

1. The top data block is completed with the following information:
   a. Name of individual completing the PICA (for example, the Process Improvement Coordinator)
   b. PICA number

    c. Date
    d. Organization and phone extension
2. The block entitled "Recommended Process Improvement" is completed with:
    a. A description of the Process Improvement
    b. Estimated labor costs associated with the new procedures
    c. Estimated nonlabor costs associated with the new procedures
    d. Any necessary one-time implementation (set-up) costs
    These costs are totaled to yield the total costs of the Corrective Action. These costs can be expressed in any number of ways, including annual costs, monthly costs, costs per manufactured item, etc. What is important however, is that all PICAs be consistent in the manner in which the costs are estimated.
3. In a similar fashion, the costs of superseded functions are calculated and totaled.
4. The fourth step is computing the difference in costs between the new procedure and the one superseded:
    a. The new method will cost $5000 more each month.
    b. In the case of a recommended change to improve a process, it could be documented that the new procedure results in less cost than the old; in other words, savings will result from implementation.

After PICAs are completed for each Process Improvement:

1. Process Improvement #2, for example, could show that an additional cost of $8000 is estimated.
2. A comparison is made between the estimated costs for each potential Process Improvement, and the most cost-effective action is selected for implementation.
3. A copy of the PICA for the selected Process Improvement is forwarded to the finance administration organization.
4. PICAs related to the FN are attached and filed for future reference.

Process Improvement Follow-Up Plan

A Process Improvement Follow-Up Plan (Figure 11.4) is necessary to continue with the financial focus. The purpose of the plan is to increase profitability in three ways:

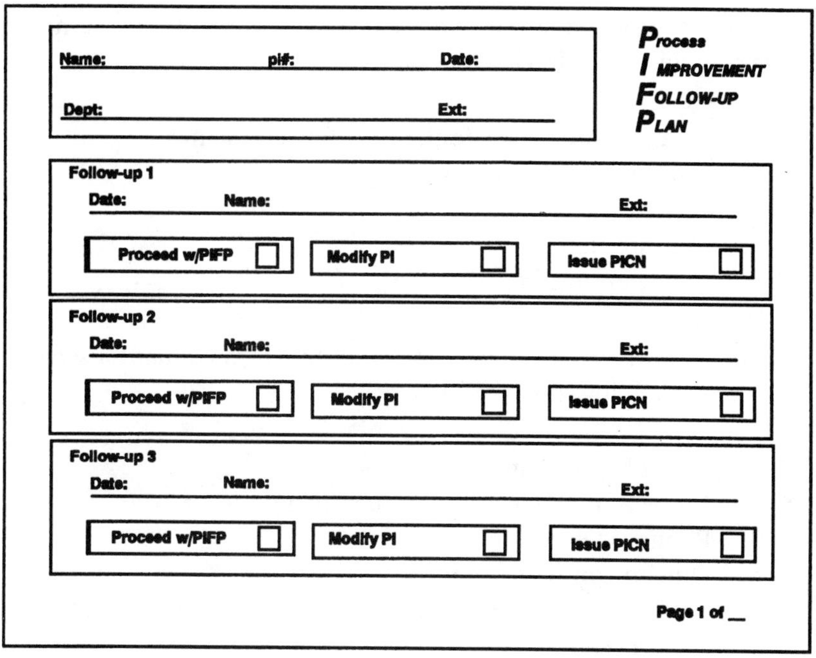

Figure 11.4. Process Improvement Follow-Up Plan

1. If the Process Improvement (Corrective Action) did not fully satisfy the cause of failure, follow-up action in a timely manner can lead to the generation of alternate procedures.

2. If a Process Improvement has proven to be successful, follow-up can alert quality engineers and other Process Improvement Coordinators that this may be useful in other applications.

3. By following up, it can be learned that the Process Improvement is no longer required. (For example, 100% inspection may have been implemented when parts coming from suppliers were generally poorly produced. The supplier may have implemented Process Improvements and corrected component problems. Receiving inspection may revert from 100% to more cost-effective sample inspection.)

The follow-up plan contains narrative explaining exactly what tests or procedures are necessary to ensure that the implemented Process Improvement has been successful. The recommended duration between subsequent follow-ups is also provided.

The first page of the plan is presented in Figure 11.4. This summary page contains the following blocks:

1. Name of supervisor responsible for follow-up
2. Process Improvement number
3. Date
4. Organization
5. Phone extension

Multiple follow-ups may be desired, and the form contains a separate area for each follow-up as follows:

1. Date of follow-up
2. Name of individual performing follow-up
3. Phone extension
4. Outcome of follow-up:
   a. The follow-up was successful, so move on to next follow-up procedure.
   b. The follow-up found anomalies; recommend revising Corrective Action.
   c. The final follow-up activity has been completed successfully, in which case, the Process Improvement Closure Notice should be issued.

Process Improvement Closure Notice

When it has been clearly determined that a Corrective Action has been successful — that the cause of the failure has been corrected — the Process Improvement Coordinator issues a Process Improvement Closure Notice (Figure 11.5). And it does just that! The Closure Notice completes the Process Improvement cycle by closing activity initiated by the Failure Notice.

Copies of the Closure Notice are forwarded to the financial administration organization and the performing organizations to alert them of the final procedural changes (if any). The data are also entered into the database for referral by all who have access. The data contained on the Closure Notice include:

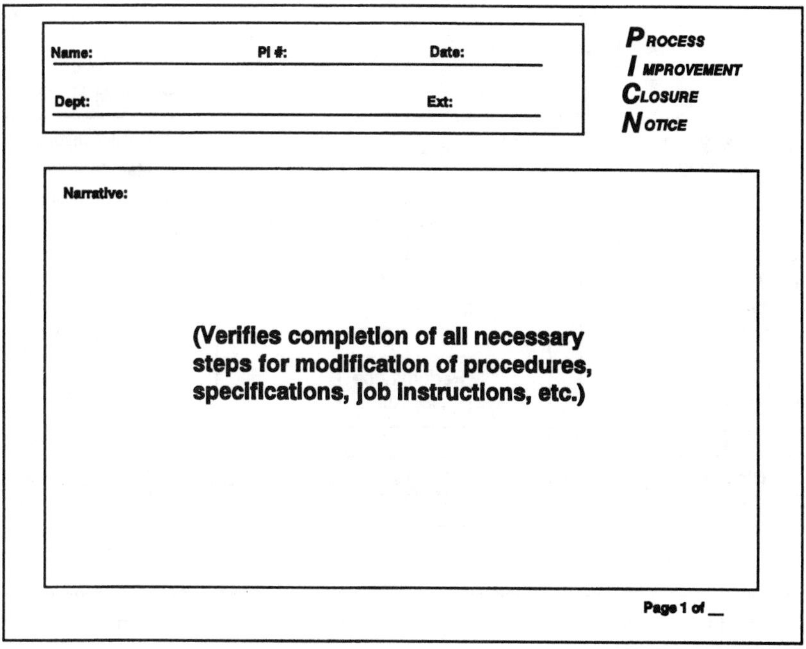

Name:	PI #:	Date:	**P**ROCESS
Dept:		Ext:	**I** MPROVEMENT
			CLOSURE
			N OTICE

Narrative:

(Verifies completion of all necessary steps for modification of procedures, specifications, job instructions, etc.)

Page 1 of __

Figure 11.5. Process Improvement Closure Notice (PICN)

1. Name
2. Process Improvement number
3. Date
4. Organization
5. Phone extension
6. Narrative explaining events performed to close the action (for example, modifying procedures)

Once the action is closed, it is no longer a Process Improvement or a Corrective Action, but becomes a standard operating procedure. Reference to the FN may also be made in this narrative portion.

Financial Administration

Financial administration (Figure 11.6) plays a critical role in Financially Focused Quality. The traditional Finance functions affected by FFQ procedures include:

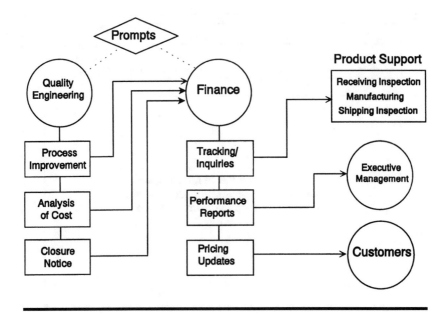

Figure 11.6. Financially Focused Quality: Financial Administration

1. Proposing/negotiating new prices and contracts
2. Tracking company performance to proposed/negotiated prices and funding

These activities are almost always reactive in nature. Within FFQ, Finance responsibilities are dramatically increased to include the following:

1. Receiving the PICA — Often, Finance will be involved in preparation of the cost analysis. Upon its receipt, Finance:
    a. Validates the data (e.g., labor rates, computer costs).
    b. Incorporates these increased costs into prices, in-process proposals, and budgets.
    c. Monitors the actual expenses incurred by organizations performing the Process Improvement procedures to verify that costs are being incurred as projected. Identifying anomalies here may lead to the realization that the designated Process Improvement procedures are not being followed properly.
2. Receives the follow-up plan, which is occasionally referred to as a tickler file, reminding Finance to prompt the Process

Improvement Coordinator periodically to ensure that follow-up actions are being taken.

3. Receives the Closure Notice. Upon receiving the Closure Notice, further justification is available for supporting negotiations of new contracts or recommending new pricing policies. Also, this alerts Finance that the changes to procedures have been formally approved and that any further changes noted while monitoring and tracking actual costs could signal other deviations.

The involvement of financial administration in the Process Improvement cycle provides executive management as well as its customers with an immediate source of critical cost data. Prior to FFQ, these data were not available until many months or sometimes years after significant changes had taken place.

When pricing, proposing, or negotiating new products or contracts, Finance relies very heavily on historical data. It is advantageous to alert Finance to the cost impact of new Process Improvements. Finance can immediately update prices or quotes to ensure that any additional effort is included.

Finance is also in the ideal position to prompt Process Improvement Coordinators (for example, Quality Engineering management) when expenditures are not in line with the PICA. Perhaps the performing organization has misunderstood the Process Improvement and is not performing it correctly. Finance, in reviewing cost data, can identify this situation and also work with the coordinator to determine the cost effectiveness of the Corrective Action.

Finally, Finance should ensure that the coordinator is adhering to the Process Improvement Follow-Up Plan, issuing alternative Corrective Actions and Closure Notices as required and in a timely manner.

Product Support Organizations

The product support organizations are those with either direct or indirect contact with the product. They often identify failures, cause failures, and correct failures. Figure 11.7 presents the role of the product support function in Financially Focused Quality:

1. Product support performance is tracked by Finance.

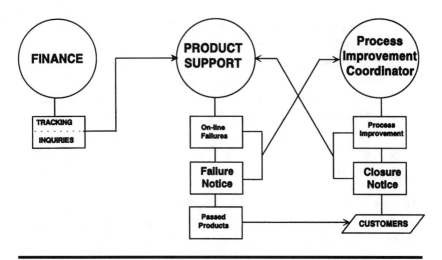

Figure 11.7. Product Support

2. When significant variances exist between actual costs and expected costs, the Finance department seeks justification from product support organizations.
3. When product support organizations identify failures, the FN is completed and forwarded to the Process Improvement Coordinator.
4. Product support organizations implement Process Improvements and Corrective Actions as assigned by the Process Improvement Coordinator. If failures continue, product support organizations originate and forward more FNs.
5. Product support organizations formalize, modify, or discontinue Process Improvements and Corrective Actions upon receipt of the Closure Notice.
6. Product support responds to inquiries from Finance.

Summary

Financially Focused Quality should be applied when an enterprise is considering taking significant steps to improve profitability. Often, on the surface, it may appear that certain activities will yield improvements in operations. The effective utilization of FFQ has revealed in many instances that costs will outweigh the benefits, and often alternative, more advantageous improvements have been implemented as a result. The key is understanding all financial ramifications.

Self-Study/Discussion Questions

1. Select a recent improvement implemented at your company. Make assumptions, and attempt to calculate the one-time implementation cost, the ongoing cost of new processes, and the costs associated with superseded activities. Summarize your findings.
2. Offer suggestions for calculating increased training costs associated with headcount reductions. What sort of cost impact could you associate with a reduction in quality?
3. In what situation would brainstorming be appropriate? When would it not be appropriate?
4. Share your experiences with brainstorming. What worked? What didn't?

Reference

DeWar, D.L., *The Quality Circle Guide to Participation Management*, Prentice-Hall, Englewood Cliffs, NJ, 1982.

Case Study A: Perky Pets — Commercial Manufacturing

A lthough the Perky Pets case study is fictional, the study is based on real-life manufacturing goals and objectives and offers valuable insight into the application of Financially Focused Quality.

The Company

Perky Pets, Inc., is a company that started out as a family business. Joe, the father, had recently retired and taken a vacation trip to Antarctica. During this trip, he became acquainted with several penguins and made the following observations:

1. Penguins are fairly intelligent.
2. Penguins have webbed feet.
3. Penguins cannot fly.
4. Penguins have scale-like, barbless feathers.
5. Penguins have flipper-like wings.
6. Penguins are likable.
7. Penguins are perky.

"Penguins are adorable!" thought Joe. Whenever he saw one, he wanted to pick one up (a small one) and squeeze it. "What an idea for a product," Joe mused. "Everyone will want one!" Joe met with his sons and together they formed Perky Pets, Inc.

The Product

Penguins have many unique properties to be considered during the product development phase. A major physical limitation the penguin product imposes is the fact that penguins require cool climates to survive. The Perky Pets engineering team developed a state-of-the-art, helmet-type cooling device that straps onto the penguin's head. The total product is the penguin and helmet, which will be sold as a unit called "Penguin-in-a-Helmet". The product will be marketed as an impulse item and sold at grocery stores throughout the United States and Canada.

Manufacturing Process

Fabrication and shipping of the cooling helmets require the following steps (see Figure A.1).

1. Various suppliers provide Perky Pets with both penguin and non-penguin manufacturing components. Penguins are obtained in the most humane manner possible. Effort is made to ensure that penguins are collected in family units which helps alleviate the trauma that could be experienced, particularly in the case of young penguins that could be separated from their parents. Research has suggested that when they are able to avoid trauma such as parental separation, penguins are more likely to maintain their perkiness.
2. Components are inspected upon receipt via appropriate sampling plans. Non-penguin components that pass receiving inspection are forwarded to the manufacturing areas. Penguins are placed in specially designed penguin quarters, where trained penguin counselors work to ensure that the penguins remain healthy and perky. Components that do not meet inspection criteria are returned to vendors.
3. Non-penguin components are then used to assemble the helmets.
4. The final manufacturing step is strapping the helmet onto the penguin, ensuring a snug but comfortable fit.
5. In the shipping area, final test and check-out are performed.
6. Cases of Penguin-in-a-Helmet are shipped to the stores, where it is hoped that anxiously awaiting customers will make purchases.

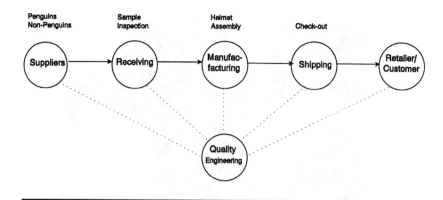

Figure A.1. Penguin-in-a-Helmet Manufacturing Process

7. The Quality Engineering organization is actively involved with each step by:
    a. Developing and issuing specifications to the supplier and receiving department.
    b. Assisting in preparation of in-process manufacturing and inspection instructions.
    c. Providing guidance for those performing final checkout, packaging, and shipping of Penguins-in-a-Helmet.

Marketing

The successful sale of the product is dependent upon an emotional response to the perky personality of the penguins. Major marketing messages have been designed to present the picture of people enjoying a partnership with the perky pet. The media plan is fairly straightforward:

1. Advertisements have been run in many magazines targeting grocery stores.
2. Direct mailings were sent to major grocery store chains throughout the U.S. and Canada.
3. Sample products have been sent to magazines and newspapers throughout the sales region, and favorable reviews have been given.
4. Press releases have also been sent to magazines and newspapers.
5. Representatives of Perky Pets' executive board have participated in the following activities:

Figure A.2. Penguin-in-a-Helmet Happy Customers

    a.   Demonstrating the product at grocery store conventions.
    b.   Making guest appearances on popular television and radio talk shows.
6.   The Perky Pets Web site has become extremely popular with Internet surfers. Many links have been established, and the site had over 200,000 hits in its first month of operation.

First and Second Weeks of Sales

What was the result of this extensive marketing plan? *Penguin-in-a-Helmet was an overnight success!*

As shown in Figure A.2, eager buyers were storming grocery stores to buy their own Penguins-in-a-Helmet. The customers were happy, grocery store management was happy, and Perky Pets, Inc., was over-joyed. Perky Pets made the cover of *Time* magazine. There were rumors in *The Wall Street Journal* that Perky Pets was planning to go public. Orders for Penguins-in-a-Helmet increased tenfold in the first week. The future indeed seemed rosy.

Third Week of Sales

During the third week, another phenomenon materialized. The encouraging start was about to be slowed substantially by an obstacle — a rather significant obstacle. Perky Pets began to experience a failure mode.

Figure A.3. Penguin-in-a-Helmet Failure Mode

Failure Mode

Sales slowed considerably, and Perky Pets had to establish a Warranty Service organization. The penguins stopped behaving in a perky manner. They became lethargic, almost catatonic. After only a few of days of perkiness, the penguins entered into a stupor.

"Either fix this penguin or give us our money back!" was the jeering cry of dissatisfied Penguin-in-a-Helmet customers across the continental U.S. and Canada. It is clear by the expression of the Perky Pets Warranty Service representative, portrayed in Figure A.3, that the company had to take immediate corrective action.

Financially Focused Quality

Fortunately for Perky Pets, Inc., the executive board had already embraced the concepts of FFQ, and, as presented in the overview in Figure A.4, they used it to incorporate a financial viewpoint in the Corrective Action process. As described below, a financial viewpoint is important from identification of the failure through to closure of Quality Engineering Corrective Actions affecting the manufacturing process.

Failure Identifiers

In this case, failures are obviously identified by the customers. Can't you just hear it? A little boy cries out to his mother, who is busy in the next

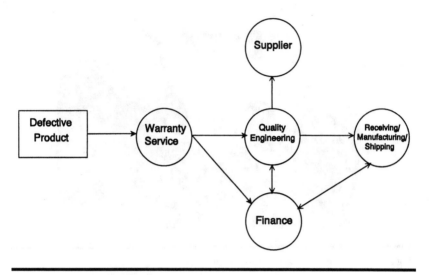

Figure A.4. Financially Focused Quality: Penguin-in-a-Helmet

room: "Mommy! This penguin isn't working right! It's broken! Can you fix it?"

The mother rushes in and sees the penguin lying prostrate on its back. Its wide, unblinking eyes are glazed, its body motionless. The lad buries his tearful face in his mother's embrace. "Oh, Mommy," he sobs. "Can you fix it?"

The mother picks up the failed product and drives to the Penguin-in-a-Helmet Warranty Service organization. Upon her arrival she is asked to take a number from the little machine. It seems there are quite a few people in line ahead of her.

The first individuals in the company to be designated as Failure Identifiers are the Warranty Service representatives, who do the following tasks:

1.   Complete the Failure Notice (see below).
2.   Attach the Failure Notice to the defective Penguin-in-a-Helmet and forward it to the Quality Engineering organization, which is designated as the Process Improvement Coordinator.

The above procedures are fairly routine for any situation, regardless of the company's Corrective Action procedure. Financially Focused Quality, however, requires the Warranty Service organization to perform the following added function:

3.   Forward a copy of the Failure Notice to the Finance organization.

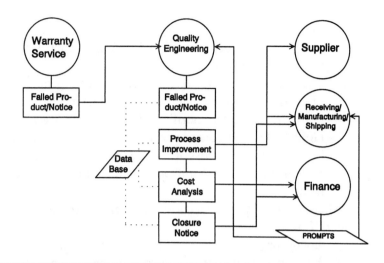

Figure A.5. Quality Engineering as Process Improvement Coordinator

Failure Notice

When filling out the Failure Notice, special attention should be given to the description of failure. Warranty Service representatives do not have formal veterinary training, and as such may not be able to write a medically accurate description of the failure. It is important, though, that the Failure Notice contain a reasonable assessment of the symptoms, as follows:

> "The Penguin-in-a-Helmet discontinued normal operation. Not only did it stop being perky, it stopped reacting to external stimuli. It ceased to eat, sleep, or breathe. Cause: unknown."

Quality Engineering

Quality Engineering functions as the Process Improvement Coordinator and (as illustrated in Figure A.5):

1. Receives the failed Penguin-in-a-Helmet.
2. Receives the Failure Notice.
3. Performs the requisite analyses and makes contact with other failure analysts, as needed, to determine the cause of penguin failure:

a. The penguins failed due to heat exposure.
b. The penguins failed because the cooling helmets were not functioning properly.
c. The cooling helmets were not functioning properly because: (i) the gas used in the helmet cooling mechanism had not been adequately purified; (ii) after a few weeks, contaminated gas clogged the expansion valve, resulting in the loss of cooling; (iii) loss of cooling resulted in an unacceptable increase in penguin body temperature, causing the failure.

4. Using a financially focused approach, analyses are performed to identify which actions are the most cost effective, and the product support organizations are notified to perform the following:
   a. A supplier representative relocates to the city where the gas supplier's factory is located. The representative works to help improve the supplier's gas purification process.
   b. Inspectors on the loading dock are required to tighten inspection criteria on all deliveries, including an increase to 100% inspection of gas.
   c. Additionally, because management is concerned about unfavorable publicity and threats from several animal rights groups, inspectors on the loading dock are also required to perform 100% examination of penguins to ensure each one is perky upon arrival, as illustrated in Figure A.6.

The following activities are required to ensure a financial focus:

1. Process Improvement Cost Analyses are generated in the Corrective Action selection process. The cost analyses for implemented Corrective Actions and Process Improvements are forwarded to Finance.
2. The cost analyses quantify the increased costs by contrasting the before and after costs:
   a. The costs associated with 100% inspection of gas vs. the prior costs of sampling
   b. The costs associated with 100% examination of penguins vs. the prior costs of sampling
   c. The costs of having a supplier representative work with the gas supplier (previously this cost did not exist)

Figure A.6. Increased Inspection

3. A Process Improvement Follow-Up Plan is developed to perform the following checks after one month from the implementation date:
   a. If receiving inspection finds that the gas is consistently meeting the increased purification standards, the on-site supplier representative may relocate back to Perky Pets' international headquarters.
   b. Also, if the gas is consistently meeting the increased purification standards, receiving inspectors may revert to the less costly technique of sampling.
   c. Similarly, if pressure from animal rights groups drops to an acceptable level, the 100% penguin examination requirement may be lifted, with a reversion to sampling.
4. Finally, when the failure mode has been corrected, Quality Engineering issues the Process Improvement Closure Notice, which formally ends the Process Improvement cycle and ensures that unnecessary process changes are discontinued.
5. The Closure Notice is forwarded to performing organizations.
6. The Closure Notice is forwarded to Finance.

Financial Administration

Finance performs the routine financial functions of any manufacturing concern, but in addition has the financially focused responsibilities shown in Figure A.7:

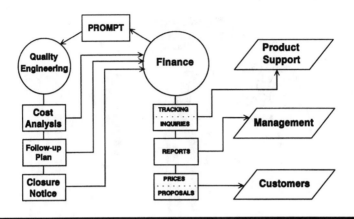

Figure A.7. Financial Administration

1. Receives and logs in the Failure Notice, awaiting Corrective Action input.
2. Receives the Process Improvement Cost Analyses for implemented Process Improvements.
3. Monitors actual costs incurred by performing organization to ensure that they are in line with the anticipated changes in costs.
4. When actual costs vary from what is anticipated based on the cost analyses, Finance investigates the variance with performing organizations to bring discrepancies to management attention.
5. Cost changes resulting from the Corrective Action are incorporated to update budgets and prices as follows:
   a. Prices are updated to reflect cost increases. The price for which a product sells must be high enough to at least allow the manufacturer to break even (unless the company is being operated as a tax shelter). To ensure that an adequate profit margin exists, Finance considers immediate price increases to offset increased production costs.
   b. Organization budgets are adjusted for the increased scope of work mandated by Corrective Actions imposed by Quality Engineering. In times of extremely tight budgets, management may expect a department to perform added requirements with existing resources. This often requires creative utilization of manpower.
6. Finance receives the Process Improvement Follow-Up Plan, and prompts Quality Engineering to ensure it is carried out as planned.

7. Finance eventually receives the Process Improvement Closure Notice. Upon receipt of the Closure Notice, Finance:
   a. Monitors actual costs incurred by performing organizations to ensure the discontinuance of any Corrective Actions that are no longer required (e.g., 100% inspection of gas).
   b. Files the Closure Notice, Follow-Up Plan, and cost analyses for future reference.
8. Finance continues to monitor, track, and prompt Quality Engineering on open Corrective Actions.

Product Support Organizations

The product support organizations perform the following functions for failures in which a financial focus is used:

1. Observe failures and complete Failure Notices.
2. Forward Failure Notices and failed hardware or penguins to Quality Engineering.
3. Perform Corrective Actions and Process Improvements as designated by Quality Engineering.
4. Modify, reduce, or discontinue Corrective Actions and Process Improvements as specified on the Process Improvement Closure Notice.

The Supplier of the Contaminated Gas

The supplier has the responsibility to meet the contractual specifications for gas purity. Perky Pets, Inc., sent a supplier representative to assist the supplier with process improvement. The process changes that take place may result in an increased cost to deliver the gas to Perky Pets. This being the case, Perky Pets is hoping that the gas supplier does not use FFQ tools in its process improvement procedures, because, if it does, any increased cost could be immediately passed on to Perky Pets. Without FFQ, the supplier may not recognize a need to raise prices for quite a while.

Perky Pets Summary

The benefits of Financially Focused Quality are clearly presented in this model. They are summarized below:

1. A high-quality Corrective Action plan is implemented, leading to timely recovery from the failure mode.
2. The selected Corrective Actions were also the most cost effective available.
3. The company was immediately able to analyze impacts to production costs, providing the opportunity to adjust prices if appropriate.
4. A Follow-Up Plan was established to ensure that the Corrective Action had been effective, providing the following benefits:
   a. If Process Improvements are not effective, immediate recognition of this fact enables Quality Engineering to generate alternative steps in a timely manner.
   b. If Process Improvements are successful, imposed steps can be reconsidered and, if possible, their magnitude reduced or eliminated. In this case, for example, the effort of working with the supplier to improve his gas purification system was successful, and 100% inspection of gas was no longer required. If it is no longer needed, it should not be done.
   c. The Finance organization provides added impetus for Quality Engineering to perform the follow-up function. Finance keeps a log of open Corrective Actions, and continues to prompt Quality Engineering for Closure Notices.

Self-Study/Discussion Questions

1. What could be the Vision Statement for Perky Pets? Mission Statement?
2. Evaluate the performance of Perky Pets in the following areas:
   a. Product development
   b. Manufacturing process
   c. Supplier relations
   d. Marketing
   e. Warranty service
3. What similar products can you imagine being potential moneymakers for Perky Pets? What problems might be encountered, and how could FFQ be used to resolve them?

Case Study B: Financially Focused Quality Implemented in Software Engineering*

Introduction

This factual case study illustrates the application of Financially Focused Quality in a major software development project. In modern software development, the process for implementing changes to the baseline configuration is made via a Problem Reporting Corrective Action System. All changes to the items in a configuration baseline are identified on a Problem/Change Report and are processed through this system. All the concepts of Financially Focused Quality are exhibited in this process.

On the AN/ALR-67 Project at Applied Technology, a division of Litton Systems, Inc., a Problem Reporting Corrective Action System is in place. This method, also known as the Problem/Change Report process, is used for all changes in requirements, problems, or enhancements to AN/ALR-67 capabilities. By adapting the Financially Focused Quality process, substantial improvements to the Problem/Change Report process were achieved.

---

* This case study is based on the technical paper "Profitable Software Quality", presented at the First World Congress on Software Quality held in San Francisco, CA, in 1995. The paper was written and presented by Robert J. Herbert, Litton Applied Technology, and Thomas M. Cappels, Lockheed Martin Missiles & Space.

This case study reviews related financially focused concepts and demonstrates how they have been applied to the AN/ALR-67 project. The resulting Process Improvements effectively transform the existing Problem/Change Report process into a software FFQ system, saving millions of dollars while maintaining high quality standards.

Financially Focused Quality Overview

Financially Focused Quality focuses on the financial implications of every function involved in delivering products. This overview summarizes those FFQ components that are now or potentially could be incorporated into Litton's Applied Technology Division. The three FFQ components are

1. Identification of the need for FFQ
2. Coordination, analysis, and selection of the new process or Corrective Action
3. Follow-up and closure of the need identified in the first component

Identification of Need

Financially Focused Quality (FFQ) begins with identification of any situation for which there is the potential for improvement (e.g., change in customer requirements, failures). Note that all employees/customers are encouraged to initiate the FFQ cycle. FFQ tools and techniques assist by ensuring a financial perspective is applied to the questions that arise at this stage of the cycle:

1. Is the need identified truly needed and/or wanted?
2. Is the current process cost effective?
3. Do employees feel all their activities contribute favorably to the product?

Coordinate/Analyze/Select Process/Corrective Action

Financially Focused Quality requires that process improvement coordination be performed by someone with a true financial perspective. In software engineering, it is usually not practical to include in the process an actual member of the financial community, so often a member of management coordinates these activities and ensures that the appropriate employees have a general understanding of the financial considerations of corrective action. An effective tool for providing such information is the training recommended by Financially Focused Quality.

As part of the routine coordination process, experts are consulted as options are generated, and eventually a revised process or Corrective Action is selected. The FFQ approach ensures that:

1. It is even practical to pursue the need identified in the first stage.
2. The appropriate amount of resources are expended in understanding the process with effective meetings and management involvement.
3. A comprehensive cost analysis is generated for those options with merit.
4. Cost is given appropriate weight in the final selection of process change or Corrective Action.

Follow-Up and Closure

The FFQ process ensures effectiveness, calling for revision, modification, and formalization of procedures to enhance profitability, thus:

1. The process ensures proper implementation of selected Corrective Actions and procedures.
2. A variance reporting mechanism allows effective feedback on unanticipated costs.
3. Software Engineering easily communicates cost data to the financial community for timely updates to proposals and pricing.
4. Selected Corrective Actions and Process Improvements are re-examined after the fact for elimination or reduction in scope to ensure every activity yields full benefit.

Background of Project to which FFQ Has Been Applied

The AN/ALR-67 is a Countermeasures Subsystem that provides Navy pilots situational awareness about the hostile and friendly radars in the environment. The AN/ALR-67 system is hosted on the F/A-18, AV-8B,

A-6E, and F-14 aircraft. The AN/ALR-67 system's software is a highly complex, real-time, embedded software package that requires quality and reliability. These high standards of quality and reliability demand that the change control process for the AN/ALR-67 system also be of superior quality. With shrinking Department of Defense budgets, it was apparent that, while quality is important, it must be balanced with cost. Litton's Advanced Technology Division adopted the FFQ philosophy of high quality at minimum costs.

Advanced Technology Division's Problem Reporting Corrective Action System

A Software Configuration Control Board controls the Advanced Technology Division's Problem/Change Report process. The Software Configuration Control Board consists of the Project's Lead Software Engineer, the Project's Lead System Engineer, a Software Quality Engineer, Project Management, and a Software Configuration Management Engineer. Throughout the Problem/Change Report process, the Software Configuration Control Board integrates the concepts of FFQ. The following is a description of the Advanced Technology Division Problem/Change Report process. Figure B.1 charts the flow of configuration control for this process.

Process Steps

1. An originator identifies a change, problem, or enhancement on a Problem/Change Report form and submits the Problem/Change Report along with supporting documentation to Software Configuration Management. Software Configuration Management reviews the Problem/Change Report for completeness, assigns a Problem/Change Report number, enters the Problem/Change Report information into a log, and distributes the agenda for an Software Configuration Control Board meeting.
2. During the Software Configuration Control Board meeting, the Software Configuration Control Board Chairperson:
    a. Consciously performs cost tradeoffs on the Problem/Change Report impact.
    b. Redirects the non-software Problem/Change Reports.

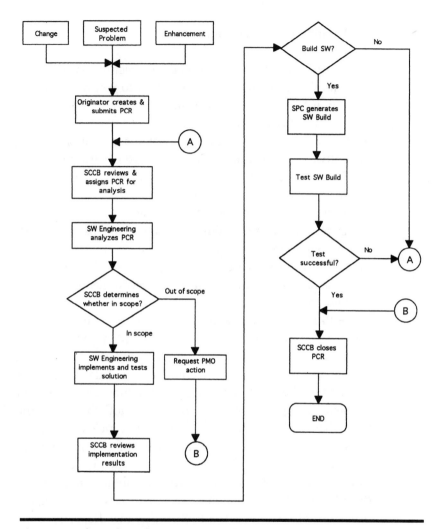

Figure B.1. Flow of Configuration Control

    c. Screens for duplication.

    d. Decides priority and suspense dates.

    e. Determines if the Problem/Change Report impacts the current product baseline, and may assign an analyst to the Problem/Change Report.

  3.   Assuming the Problem/Change Report is assigned, the analyst:

    a. Assesses the problem, as well as the costs and risks for the recommended solution.

    b. Looks at interim fixes or possible work-arounds.

    c.　Explores alternative solutions.

    d.　Recommends solutions.

    e.　Identifies all affected areas on the Problem/Change Report.

    f.　Submits the Problem/Change Report to Software Configuration Management to schedule a Software Configuration Control Board Analysis Review.

4.　At the Software Configuration Control Board Analysis Review, the Software Configuration Control Board Chairperson:

    a.　Reviews the Problem/Change Report analysis and recommended solutions.

    b.　Approves or disapproves the analysis.

    c.　Determines whether additional walk-throughs and/or testing requirements are required.

    d.　Coordinates with areas identified.

    e.　Schedules implementation of the recommended changes and assigns an Implementor.

5.　If the Problem/Change Report is out of scope, the Problem/Change Report is submitted to Program Management Office for disposition. The Software Configuration Control Board will initiate an Engineering Change Request or Order for a Problem/Change Report that impacts the product baseline(s). The Problem/Change Report will remain open until the Configuration Change Control Board returns the Engineering Change Request or Order disposition. Software Configuration management documents the disposition/assignment in the Software Configuration Control Board meeting minutes and updates their Problem/Change Report log. If the Software Configuration Control Board rejects a Problem/Change Report, it is returned with comments to the originator and closed by Software Configuration management.

6.　If the Problem/Change Report is in scope, the assigned SW engineer:

    a.　Implements the recommended solution per schedule.

    b.　Performs all minimum amount of testing and notifies Software Configuration management that the Problem/Change Report is ready for the Software Configuration Control Board Implementation Review.

7.　At the Software Configuration Control Board Implementation Review, the chairman of the Software Configuration Control Board:

    a. Reviews a version difference report by comparing changed files to the baseline.

    b. Reviews the Problem/Change Report with the version difference report.

    c. Ensures that the appropriate portions of the Software Development folder have been updated with the change files. If all the items are complete, the chairman submits the folder and attachments to Software Configuration management for incorporation into the developmental configuration and authorizes a Software Build to be made.

8. If the Software Build is not authorized, the Software Configuration Control Board returns the Software Development folder, Problem/Change Report, and supporting documentation to the Implementor for Corrective Action.

9. If the Software Build is authorized, Software Configuration management incorporates the changed files into the baseline and generates a Software Build.

10. The Implementor installs the Software Build while being witnessed by a Software Quality Engineer and performs appropriate software qualification tests.

11. If the testing is not successful, the Implementor returns the Problem/Change Report and support documentation back to the chairman of the Software Configuration Control Board for Corrective Action and disposition. If new problems are encountered during the testing, Problem/Change Reports are generated and submitted to Software Configuration management.

12. If the test is successful, Software Configuration management prepares the Problem/Change Report for Software Configuration Control Board Closure Review. The chairman of the Software Configuration Control Board reviews the recommended closure report and authorizes the Problem/Change Report closure contained in the report. The Software Configuration Control Board then reviews, signs, and authorizes closure or approval of the Problem/Change Report. Software Configuration Management updates the Problem/Change Report log, files the closed Problem/Change Reports in the Software Development Library, and documents them in the Software Configuration Control Board minutes.

Application of FFQ to the Problem/Change Report Process

Litton's Applied Technology Division's Software Development organization successfully applied FFQ to their Problem/Change Report process. This section provides an in-depth explanation of how this was achieved in three FFQ stages: (1) identification of the need for FFQ; (2) coordination, analysis, and selection of the new process; and (3) follow-up and closure of the need identified.

Need for Financially Focused Quality Identified

In this implementation, management in Litton's Applied Technology Division's Software Development organization first familiarized itself with FFQ by undergoing Financially Focus Quality training. This training was accomplished in two ways. First, employees met with FFQ practitioners to gain an understanding of FFQ tools and techniques. Second, FFQ publications were studied to gain a basic understanding for financial concepts and to learn about FFQ case studies that had yielded substantial savings.

Litton management recognized many similarities between several case studies and the operations in the Software Development organization. Recognizing the potential value of applying FFQ concepts and techniques, management asked its engineers to review each step of the Problem/Change Report process. At each step, employees were to question the cost effectiveness of the process and to attempt to validate that all their activities contributed favorably to their software product. The responses confirmed management's belief that the potential existed for Process Improvement in the Problem/Change Report process. Thus, stage two of the FFQ process began.

Coordination, Analysis, and Selection of Process Improvement

The Lead Software Engineer was assigned responsibility as the FFQ Coordinator. As such, he was tasked with determining the causal/beneficial relationship each procedure made to the bottom line. The goal was to determine whether steps could be eliminated which would reduce cost and still maintain quality.

Previously in the process, Litton was using a form called a Secondary Problem/Change Report. This was another Problem/Change Report that was written at the completion of the analyst's review of the Problem/Change Report and prior to the Software Configuration Control Board Analysis Review. This additional form was allocated for each affected computer software component identified by the analyst. The idea was that these Secondary Problem/Change Reports would be closed as each computer software component was corrected so changes could be submitted to the developmental configuration baseline prior to having the problem completely solved.

Financially Focused Quality algorithms led to significant improvements in and substantial savings for the Problem/Change Report process. To derive the data for this FFQ analysis, a financial analyst either performed time studies or queried company databases to determine a historical basis for average hours spent for each Problem/Change Report and for each Secondary Problem/Change Report.

The analyst also required access to data allowing determination of the average direct labor rate of employees administering the reports. Finally, factors were determined to ensure that appropriate allocated costs were included for generating an overall average hourly rate. Such costs may include overhead, general and administrative costs, and pooled direct costs such as computer systems or common services.

The total annual cost of Problem/Change Report processing was calculated at $14.7 million dollars. And, yes, Litton's process was critical to maintaining high quality and reliability in their software, but FFQ was called on to see if the same high levels of quality could be maintained at a lower cost. The vast majority of costs resulted from the administration of the Secondary Problem/Change Reports. After several weeks of close study by the engineers and management in the group, a few changes were made in the processing of the Problem/Change Report, and the Secondary Problem/Change Report was totally eliminated from the process!

Calculations proved that the "after-FFQ cost" was only $3.5 million, or a savings of over $11 million resulting from discontinuing Secondary Process/Change Report processing. Also, even after including a few additional steps in Problem/Change Report processing, the overall impact to the time required for Problem/Change Report processing dropped, because there was no longer a need to track individual computer software components and the corresponding Secondary Problem/Change Report.

Follow-Up and Closure

After a few months of trying the modified process, it was formally recognized that the Secondary Problem/Change Report processing was indeed not cost effective. In fact, it was very expensive. The additional form increased the software engineer's documentation by as much as fivefold. One case was documented where five Problem/Change Reports generated an additional 25 Secondary Problem/Change Reports. In addition, Litton found that all the Problem/Change Reports and their Secondary Problem/Change Reports were closed at the same time. In fact, closing these Problem/Change Reports simultaneously was inevitable because of the nature of the problems the Problem/Change Reports were identifying.

Typically, if a problem affects more than one Computer Software Component, the problem lies in the communication between the components. When this occurs, the only way to determine if the problem has been corrected is to implement a complete solution and test the components together. This means that the components must be implemented, tested, and therefore submitted for closure at the same time!

The Secondary Problem/Change Report handling did nothing for the process except add cost to the product. Accordingly, the Secondary Problem/Change Report form and all associated tasks were formally deleted from written task descriptions and job instructions.

Future Improvements

Litton is continuing to seek new ways to improve its process. Currently, Litton is undertaking the complete rewrite of the AN/ALR-67 software to make it more maintainable. In the course of this development, improvement of Litton's current Problem/Change Report process was initiated. Litton is concentrating on computerizing the Problem/Change Report process. Currently, the Problem/Change Report is filled out by hand and then placed on the flat database called Filemaker Pro, which is hosted on a Macintosh computer. Potential areas of improvement and cost savings lie in how the information flows to people involved in the system. There is a lot of paper work and disconnects because individuals do not receive or lose their paper work.

Currently, Litton is implementing its Problem/Change Report process on a workstation that is connected to a dedicated local area network (LAN). The Problem/Change Reports and the forms will be on a relational database, and an event promotion system is being developed. The relational database will be used to generate metrics so Litton can begin measurement of its overall software development process.

It is envisioned that Problem/Change Reports will no longer be written by hand. Instead, when a Problem/Change Report is written, the originator will type the problem into the Problem/Change Report form on the network. The Problem/Change Report then will be sent via e-mail to Software Configuration management. Software Configuration management will then be able to generate the Software Configuration Control Board agenda on-line and notify the Software Configuration Control Board members by e-mail.

As each Problem/Change Report goes through each step of the process, it will be promoted by the on-line Problem/Change Report system to a new state, and electronic notification will be used to maintain communication with all affected parties. Litton anticipates that this system will greatly reduce the load on Software Configuration management personnel who currently are busy enough with paper work.

Conclusion

This case study has demonstrated successful FFQ application to software development, with substantial savings for Litton's Applied Technology Division. Effectively eliminating waste in business will ensure that every effort will positively impact society.

Self-Study/Discussion Questions

1. How does the FFQ approach utilized at Litton differ from the approach used in a manufacturing environment? For a service industry?
2. What circumstance, if any, could justify the continued use of Secondary Problem/Change Reports. Why do you think they were used in the first place?

Reference

Cappels, T.M., *Full-Cycle Corrective Action: Managing for Quality and Profits,* Quality Press, Milwaukee, WI, 1994.

Case Study C: Bob's Purple Bayou Café*

Introduction

Bob's Purple Bayou Café is a small, successful restaurant with take-out/delivery and catering services. The café became successful very quickly and was beginning to experience problems with its processes. The café was making a profit, but customer complaints were continuing to increase. Bob has plans to expand to other locations, but before he does, he needs to assess his current operation and adjust the processes accordingly. The employees at Bob's Purple Bayou had recently received Financially Focused Quality training and planned to use their new knowledge to benefit the café. They began by doing some financial analysis and customer surveys.

A study was performed to determine how much revenue was generated by each of the three income-producing activities. Also, customer satisfaction surveys were administered. Problems noted over a period of one month were grouped by activity with corresponding revenue data, as shown in Table C.1. After reviewing the customer complaints, Bob realized he must act quickly to improve his operation. The Process Improvement process first focused on delivery orders, because this activity contributed the most revenue and highest percentage of problems.

* This case study was developed based on a paper prepared by Mary Citrino, Elaine Ginter, Dyanne Holmes, and Carey Hood.

Table C.1. Bob's Purple Bayou Café Problems

Problem	Problems per month	% orders with problems	% of revenue
Take-out orders	1300 of 3000	43	55
Delivery orders	200 of 900	22	25
Restaurant service	40 of 1500	3	20

Current Status

Bob read several articles about Total Quality Management and other process improvement materials. At the time, he planned on going to training classes and hiring a manager. After considering the costs, Bob decided to explore other less-expensive alternatives. He stumbled on some pieces about Financially Focused Quality and found that he agreed with many of its basic tenets.

Bob scoured the library for information on Financially Focused Quality. He read much about cost-effective quality management and realized he needed to create business objectives, a Vision Statement, a Mission Statement, and a continuous improvement process. After much research, Bob brought in his wife, Thelma Lou, who had prior business management experience, to manage the café's operations, and he developed a set of business objectives.

Business Objectives

1. Delight the customer.
2. Make a profit.
3. Expand to other locations.
4. Provide superior product quality.
5. Implement just-in-time inventory.

Vision

Bob's Purple Bayou Café's vision is to "delight our customers across the nation with superior quality food and service."

Table C.2. Bob's Purple Bayou Café Processes

Process	Assigned Group
Kitchen functions	Kitchen staff
Management and staffing	Bob
Training	Bob/chef/manager
Logistics	Manager
Order processing	Counter staff
Quality control	All team members
Research competition's processes	Bob/manager/chef

Mission Statement

Bob's Purple Bayou Café's Mission Statement is divided into three elements:

1. *Customers:* "Provide our customers with high-quality products and a pleasurable dining experience."
2. *People:* "Recruit dynamic, customer-oriented employees and provide excellent training programs geared toward continuous improvement. Offer our employees opportunities for advancement, including an excellent benefits package."
3. *Communities:* "Act as a responsible business entity within the community by supporting local programs."

To address continuous improvement methods, Bob planned to implement strategy processes to address quality improvement, operational procedures, and human resource issues.

Processes To Be Improved

Table C.2 presents the seven functions being targeted for improvement, and the person or team assigned corresponding responsibility.

Expected Outcome

The expected outcome is to improve the take-out order process and to reduce customer complaints from 43% to 15% within the first 90 days.

Support Behaviors Needed by Top Management

Bob needs to be visible and involved in the improvement process. Once Bob has a good understanding of what has to be accomplished, he should bring his employees together to participate as an improvement team with the task of identifying, analyzing, and recommending new ideas for meeting the company's mission. He also needs to perform an effective team launch. This would include skills training that will foster trust and cohesiveness.

Next, the team needs to create goals and strategies to meet their business objectives. The team will create an audit improvement process measurable against company goals. Management will also need to assess the competition continuously.

Type of Team/Group To Be Utilized

The type of team that Bob will create is a Process Improvement Team. This type of team is unique in that all employees in the group must partake in the process. In a small company such as Bob's Purple Bayou, the entire staff would participate and be responsible for implementation of new plans. Management is heavily involved in this type of team and regularly attends the frequent, but short meetings. The chef will be a team leader in the food preparation process. This will result in his buy-in and force him to delegate some responsibility.

Training

Training will be on the job, led by Thelma Lou. It will include team-work-building skills such as role-playing, conflict resolution, and effective communication methods. It will also include process management and problem-solving procedures. Also included will be training for the chef and his team by the manufacturer of the new state-of-the-art equipment. Cross-training will occur where applicable.

Measurements To Be Applied

Thelma Lou, who by this time had become a strong supporter of Financially Focused Quality, convinced Bob that measurements are essential.

She said if you cannot measure it, you cannot control it. If you cannot control it, you cannot manage it. If you cannot manage it, you cannot improve it. Consequently, the improvement team established the following measurements:

1. Effectiveness
    a. Reliability
    b. Serviceability
    c. Responsiveness
    d. Appearance/image
    e. Complaints per order output
    f. Percentage of on-time deliveries
    g. Customer satisfaction
2. Efficiency
    a. Processing time per order
    b. Resources needed per order
    c. Orders per hour
3. Financials
    a. Profit percentage of sales
    b. Assets
    c. Operating costs
    d. Quarterly/annual sales
4. Business
    a. Safety
    b. Market share/competition
    c. Inventory turns
    d. Continuous improvement

Supplier Management Strategy

Bob's Purple Bayou Café realized that they needed to develop a relationship with their suppliers. Bob and Thelma Lou established a team which included the manager, the chef, and supplier representatives. They set up a process that included:

1. Just-in-time inventory
2. Automatic reorder
3. Commitments for special orders

Table C.3. Bob's Initial Report Card

Elements	Grade	Comments
1. Deliver products with outstanding quality.	D–	43% of orders result in complaints, cold food, late orders
2. React quickly in handling complaints.	D–	No service recovery process
3. Look for trends, correct problems before they become complaints.	D	No process improvement efforts in place
4. Give feedback to employees.	F	No communication or interaction

These processes will allow the café to meet their business objectives, eliminate waste, and meet their customer satisfaction requirements. The team established an audit procedure to measure supplier performance.

Customer Satisfaction Characteristics

The working team members, along with the management, continually assess the café's customer satisfaction. The characteristics revealed have been (1) high frequency of complaints, (2) cold food, and (3) late orders.

Report Cards

The team created a report card (see Table C.3) on the café's beginning efforts. In the intervening weeks, the improvement team went to work on the complaints, cold food, and late orders. Table C.4 shows the success of their efforts and teamwork after 90 days. The team continues to use the report card to measure their success to date. Bob is looking forward to achieving and maintaining straight A's.

Communications

In an ongoing effort to communicate effectively with employees, management reviews all measurements on a continuous basis. Management also involves all employees in the continuous improvement process to meet the café's goals and objectives.

Table C.4. Bob's Improved Report Card

Elements	Grade	Comments
1. Deliver products with outstanding quality.	B+	88% of orders are on time with no complaints
2. React quickly in handling complaints.	B+	Service recovery in effect 85% of the time
3. Look for trends, correct problems before they become complaints.	B	Process improvement implemented; complaints decreased 82%
4. Give feedback to employees.	A–	Continuous communication processes implemented

Feedback

Part of the café's continuous improvement includes external customer satisfaction surveys. These surveys are handed out to all customers and are included in all take-out and delivery orders. The team wanted to make the complaint process simple and have the customer feel the café would listen and act on their complaints. The café also asked for suggestions on how service could be improved. One of the elements in the survey was customer experience. In addition to a comments section, the survey included questions regarding:

1. Overall satisfaction
2. Whether the customer would buy again
3. Taste
4. Quality and presentation of food
5. On-time delivery of orders
6. Whether they would recommend the café to friends
7. Comparison to the competition
8. Friendliness of staff

Network Structure

Knowing what the disadvantages were going in, Bob and his Thelma Lou modified the case management network structure to eliminate any pitfalls. Following are the elements of this structure used by the cafe:

1. Strong customer focus
2. Flexibility
3. Total accountability
4. Coordination between functions
5. Cross-functionality
6. Responsiveness to customer needs

Conclusion

With the improvements that Bob has realized at his café, he can now focus on expanding his business. His expanded operations can be run effectively and efficiently. Bob's knowledge of finance, cost, process improvement, and customer satisfaction will enable him to be successful in his plans for future expansion.

Self-Study/Discussion Questions

1. Identify the specific Process Improvements implemented at Bob's Purple Bayou and attempt to compute (a) the cost of implementation, and (b) the cost of maintaining the process.
2. If you were Bob and you were receiving straight A's on your report card, how would you go about your plans to expand the business? What steps would you take? What type of resources would you be willing to expend? What options would you consider? How would you perform research?
3. If you were a supplier for Bob's Purple Bayou, how would you work with Bob for the mutual benefit of your businesses?

Reference

Harrington, J.H. and Harrington, J.S., *Total Improvement Management*, McGraw-Hill, New York, 1995.

Case Study D: One Company's Success With Outsourcing

Introduction

On August 30, 1994, Lockheed Corporation and Martin Marietta jointly announced that their respective boards of directors had unanimously agreed to merge the two corporations through an exchange of common stock valued in excess of $10 billion. There were millions of dollars in merger-related expenses, and the new company — Lockheed Martin — expected to be reimbursed by their government customers for such costs. To justify the merger-related expenses, Lockheed Martin had to prove that the consolidation would result in tremendous savings. For example, there would no longer be the need for each company to have its own board of directors. An entire set of executive management could be eliminated.

The merger team identified other areas of savings as well, and several Centers of Excellence (COE) were established. Each COE was an area where the economies of scale could be fully utilized and related operations would be centralized. This study focuses on the Machining Outsourcing Center of Excellence and how the application of Financially Focused Quality (FFQ) at Lockheed Martin Missiles & Space (LMMS) in Sunnyvale, CA, substantially increased profitability.

Machining Outsourcing Initiative

The Machining Outsourcing Initiative (MOI) is the end product of the Machining COE study. This study concluded that, even if all machining for the entire Space and Strategic Missiles Sector of Lockheed Martin were centralized, it would still be less expensive to outsource production-machining activity. Therefore, the MOI was chartered to provide high-quality, machined hardware at optimized cost and scheduling using the best industry resources available. With these resources, sector hardware requirements from product development through final production phases would be met via a network of internal manufacturing centers and key outside suppliers.

MOI Mission Statement

Accordingly, the MOI Team's mission was expressed as follows: "To satisfy Space and Strategic Missile Sector requirements for machined mechanical components with the best value to customers through:

- A fully implemented Sector Make-or-Buy Policy
- A standardized outsourcing process (Sector Purchasing Agreements)
- Strategic partnerships with 'Best-in-Class' suppliers
- Consolidation of in-house resources to provide a lean and agile development/modification capability."

Financially Focused Quality

Financially Focused Quality (FFQ) is a proven management system designed for maximum quality at minimum cost. Marshall Kyger (Director of LMMS Programs, Planning, and Analysis) says that FFQ "takes the philosophy of Total Quality Management one important step further by including the financial community in the process from the beginning, and integrating recognition of cost at each step of the process. This happy marriage of the finance and quality communities offers the prospect of improving the translation of better quality to the bottom line" (Cappels, 1994).

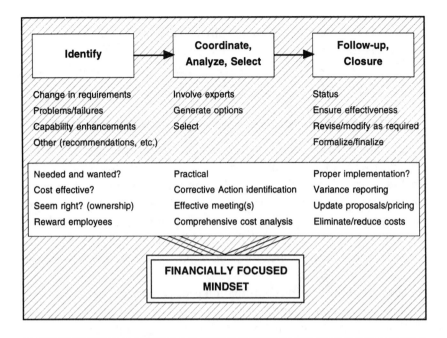

Figure D.1. Outsourcing

Financially Focused Quality Overview

Financially Focused Quality focuses on developing the Full-Cycle Mindset, which gives primary consideration to financial implications of every function involved in delivering products and services. Figure D.1 summarizes those FFQ components incorporated into the MOI activity in three stages: (1) identification of the need for FFQ; (2) coordination, analysis, and selection of the Process Improvement or Corrective Action; and (3) follow-up and closure of the need identified in (1).

Identification of Need

Financially Focused Quality begins with identification of any situation for which there is the potential for improvement (e.g., change in customer requirements, failures). Note that all employees/customers are encouraged to initiate the FFQ cycle. FFQ tools and techniques assist by ensuring that a financial perspective is applied to the options that arise at this stage of the cycle:

1. Is the need identified actually needed and/or wanted?
2. Is the current process or that proposed cost effective?
3. Do employees feel all their activities contribute favorably to the product/process?

Coordinate/Analyze/Select Process/Corrective Action

Financially Focused Quality requires that someone perform FFQ coordination with a true financial perspective. It is not always practical to include a member of the financial community, so often a member of management coordinates these activities and ensures that the appropriate employees have a general understanding of pertinent financial considerations. An effective tool for providing such information is the ASQC Quality Press publication, *Financially Focused Quality: Managing for Quality and Profits.*

As part of the routine coordination process, experts are consulted as options are generated and, eventually, a revised process or Corrective Action is selected. The FFQ approach ensures the following:

1. The practicality of pursuing the need identified in the first stage
2. Expending an appropriate amount of resources to understand the process with effective meetings and management involvement
3. Generation of comprehensive cost analyses for options with merit
4. Giving appropriate weight to cost when selecting options

Follow-Up and Closure

The FFQ process ensures effectiveness, calling for revision, modification, and formalization of procedures to enhance profitability:

1. The process ensures proper implementation of selected options.
2. A reporting mechanism allows cost tracking.
3. Cost data can be used for proposal and pricing updates.
4. Selected options are reexamined after the fact for elimination or reduction in scope to ensure that every activity yields full benefit.

The Machining Outsourcing Initiative and Financially Focused Quality

Lockheed Martin Missiles & Space has saved millions of dollars by applying the financial principles of FFQ to its manufacturing and TQM processes

(Cappels, 1997). For the MOI, the decision was made to create a full-time position of Business Deputy, who would have the following responsibilities:

1. Propose/negotiate resources.
2. Establish budgets.
3. Direct the tracking of weekly labor and nonlabor expenses.
4. Summarize and report actuals vs. budgets and analyze variances.
5. Design/implement program-unique cost models for forecasting and reporting MOI savings.
6. Establish/oversee MOI parts forecasting and tracking system.
7. Oversee MOI performance schedules/milestone tracking system.
8. Maintain the Full-Cycle Mindset, ensuring that recognition of cost is integrated at each step of the process.

The individual selected for this position had a broad, well-founded financial background and several years of direct experience working with the FFQ management system.

Cost Modeling

Financially Focused Quality concepts first were utilized to validate the belief that Lockheed Martin Missiles & Space would save significant resources by outsourcing production-machining effort. Figure D.2 is an example of the cost modeling analysis, which looked beyond the mere elimination of hourly machining personnel. Also considered were all support labor costs, overhead, and G&A, which had to be compared with the increased costs for administering the MOI. FFQ tools eased the identification and costing of such tasks as transportation, source inspection, and supplier certification.

Meeting Schedules

After the MOI received approval, the implementation process began. For the first 3 months of MOI implementation, daily team meetings were held with an average of 15 people in attendance. The Business Deputy noted that much discussion involved small groups of people, while the majority of the attendees did not participate. The deputy recommended that splinter meetings be called as needed, with the

Part Number: 5625389   MOI BUY COST MODEL

Line #	Element	$/Factor	Extended	Qty	Total /part	$	H/C
	Anticipated Quantity: 205						
	Vendor Quote per unit:	$474			$474		
A	Fixed cost per part number: RFQ Prep/coord:	2.1%	$10	205	$0	$125757	0.6
B	Variable cost/part #: RFQ/Quality/Planning/ Design/Buyer/Pull specs/Planner/M&P/ Buyer/Design Eng/Tech Interface	8.2%	$39	205	$0	$492000	3.0
C	Variable cost/total parts: Receiving/Source Inspection	3.0%			$14	$180000	1.3
	Total Contract-Direct MOI Cost				$488		4.9
D	Allocated MOI costs (ADC): /Leader/Business/M&P:	14.4%			$54	$864000	5.5
E	Unique labor costs (supply matl/equipt)		$0	205	$0	$0	0.0
E	Unique nonlabor costs (matl/equipt)			205	$0		
	Total	27.7%			$542		10.4

* 2.1% represents an AVERAGE - it's actually approx 2.0 hrs per part number.

PWO includes: Property Management, shipping, rework, (don't need full-time leader & full-time business: $200K per year for rework for which supplier does not pay.

Cost type		A %	B %	C %	D %	E %
A	RFQ coordination/prep-MOI	2.1%				
B	Buy Part Planner-MOI		0.5%			
B	Order Control-MOI		0.5%			
B	Quality Coding-MOI		0.5%			
B	Buyer		0.5%			
E	Pull material-MOI					varies
B	Pull PDSE's (MOI)		0.2%			varies
F	Transportation (see below)					
B	Request for info		2.0%			
B	Engineering		3.0%			
D	Parts status				0.4%	
C	Source			2.1%		
F	Shipment (see below)					
C	Receiving			0.3%		
C	Receiving inspection			0.3%		
D	Factory pool travel				0.5%	
D	Program Mgmt				0.7%	
D	Business Management				0.5%	
B	Production Engineering				4.0%	
C	Stocking			0.3%		
D	Eng pool travel				1.5%	
F	Freight					
D	Material adj. factor (MRB)		1.0%		1.0%	
D	Material planning				1.6%	
B	Quality Engineering				4.0%	
C	Source Planning					varies
D	Source certification				0.1%	
D	Engineering certification				0.1%	
	Total	2.1%	8.2%	3.0%	14.4%	27.7%
S/B	Rates impact (in rate)	2.1%	11.3%	3.0%	11.3%	27.7%
A	Fixed cost divided by number of parts					
B	Variable cost based on unit price					
C	Variable cost based on unit price x # of units					
D	Pool costs allocated to OS dollars					
E	fixed costs when we supply materials (Direct to contract) - varies w/Part					
F	Govt Bill of Lading: not billed to customer					

Figure D.2. Cost Modeling Process

requirement of only a weekly team meeting. The use of e-mail and the telephone greatly facilitated this Process Improvement.

With the elimination of the daily MOI meetings, the Business Deputy recognized that he could perform his assigned tasks in less than the 40 hours a week budgeted for the position. As a result, he volunteered for

...ges made the MOI more attractive to programs and MOI team ...nbers. Suppliers, too, were much happier not having to prepare ...tes for 100 parts at once — they also benefited by the load leveling.

lf-Study/Discussion Questions

1. What are some of the negative impacts of outsourcing?
2. When overhead is allocated to labor hours or labor dollars, how does outsourcing affect overhead rates?
3. What are some of the positive benefits of outsourcing?
4. What operations in your company could conceivably be outsourced? What are the pluses? The minuses?

eferences

appels, T.M., *Full-Cycle Corrective Action: Managing for Quality and Profits*, Quality Press, Milwaukee, WI, 1994.

appels, T.M., Outsourcing: one company's success story, presented at Computer Associates, Inc., Business Applications Conference, New Orleans, LA, July 13–18, 1997.

several other special assignments, allowing for a m
of assignments within the finance and business op(

Contracting Process Improvement

The Business Deputy was concerned about the prop(
lishing contracts with machined parts suppliers. The
for a 2-month contracting cycle as follows:

1. Each program provides machined parts requ
   ties and need dates) to the Business Deputy
2. The BD consolidates and forwards list to th
   Engineer (ME), who pulls specifications/dra\
   suppliers for quotes.
3. The ME forwards list to the Materials and Pr
   Engineer to ensure that the correct M&P instr
   pany the package.
4. The M&P engineer forwards the package to Qu
   (QA), who supplies quality codes and packaging
   the package.
5. QA forwards the package to the Buyer who assem
   ages and sends them to the MOI Lead Buyer in
6. The MOI Lead Buyer in Denver consolidates the
   ages with packages from other companies in th
   sends them to the supplier to for quote preparati

The Business Deputy noted the following faults in this pr(

1. Programs could not obtain machined parts in less
   months.
2. The workload required for quote package preparati(
   stable. Each person must process as many as 100 p;
   week and then have no related effort for 7 weeks.

Financially Focused Quality concepts were used in developin;
dure allowing the different MOI team members to perform tl
concurrently. This greatly reduced the stress and time required
ing quote packages. In addition, exceptions were made to the
packages could only be released to suppliers in 2-month interval

Case Study E: Hotel Operations

T here are three major components of Financially Focused Quality for hotel operations: (1) the Failure Identifier (Segment I); (2) the Management Staff Meeting (Segment II); and (3) the Hotel Controller Office (Segment III, providing the financial viewpoint).

Financially Focused Quality in Hotel Operations: Overview

Figure E.1 presents the FFQ process flow.

Failure or Opportunity Identifier (Segment I)

The FFQ cycle beings with a failure. In a hotel operation, a failure may take many forms, such as:

1. Cases of food poisoning occur in a hotel restaurant or at a banquet.
2. Power fails.
3. Guestrooms are not made up fast enough to accommodate check-in of guests.
4. Hotel guest is taken to airport in hotel courtesy car and arrives late for flight.
5. Wake-up calls are not made as promised.
6. Reservations are lost.

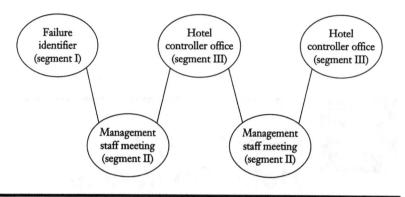

Figure E.1. Overview of Hotel Operations

7. Hotel rates are misquoted.
8. The band in the hotel lounge is too loud and keeps guests awake.
9. Appliances in hotel rooms not operating properly.
10. The list goes on and on.

The formal FFQ procedures used for a manufacturing environment includes reference to two different but similar forms, generically referred to as the Failure Notice or the Process Improvement Recommendation. Similar forms are easily adapted to hotel operations. This case study deals with a specific failure; therefore, the Failure Notice is used. There are numerous ways that failures are brought to management's attention, including the following:

1. Guest complaints and observations
2. Non-management complaints and observations
3. Management complaints and observations

When a failure is identified, the FFQ Failure Identifier (Segment I) performs these steps:

1. Generates the Failure Notice.
2. Forwards the Failure Notice for review at the weekly management staff meeting. If the failure is of significant urgency, a special meeting may be called.

Management Staff Meeting (Segment II)

The supervisors of all hotel departments attend the staff meeting. As a result, management of all potential failure solvers are in attendance. The following actions are taken in regard to the Failure Notice:

1. The notice is reviewed with attendees.
2. Potential causes are discussed.
3. Potential Corrective Actions and Process Improvements are generated.
4. Cost analyses of the Corrective Actions and Process Improvements are determined for the proposed steps.
5. Implementation of the most cost-effective Process Improvement is assigned to appropriate department supervisor.
6. A Corrective Action/Process Improvement Follow-Up Plan is developed.
7. The cost analysis corresponding to the selected Corrective Action and follow-up plan are forwarded to the Hotel Controller's department.

Hotel Controller Office (Segment III)

The hotel controller and staff provide the focus for including consideration of financial impacts with the Corrective Action cycle. Although Financially Focused Quality is a mindset which should be instilled in the minds of all company employees, the involvement of an individual specifically trained in the science of Finance is particularly beneficial. The hotel controller is such an individual.

The Hotel Controller Office performs the following tasks:

1. Attends and provides input to the management staff meeting (Segment II).
2. Updates pricing policies and budgets utilizing the cost analysis.
3. Tracks Corrective Action expenditures to the cost analysis.
4. Prompts organization supervisors to comply with the requirements of the follow-up plan.
5. Prompts the generation of the Corrective Action/Process Improvement Closure Notice.

Management Staff Meeting (Segment II)

Final actions to close the Corrective Action cycle are taken as follows:

1. When it has been determined that the Process Improvement has successfully eliminated the failure, the Closure Notice is issued to formally approve or modify Process Improvement procedures.
2. Responsibility for implementation of any Closure Notice activities is assigned to the appropriate department supervisors.

Hotel Controller Office (Segment III)

The Hotel Controller Office continues tracking costs and prompting closure of open Failure Notices.

Financially Focused Quality in the Tiara Plaza Hotel

The Tiara Plaza Hotel is a full-service hotel, with two restaurants (coffee shop and fine dining), a lounge that features live music into the wee hours of the night, several ballrooms, many meeting rooms, and 400 guest rooms. It is located in San Mateo, CA, about 10 miles south of the San Francisco International Airport.

The Front Office

Hamid manages the front office of this large, independent hotel. His department is comprised of the following functions:

1. The Front Desk (cashier and registration)
2. Reservations
3. PBX (switchboard) operations (including wake-up service)
4. Night audit (close the financial books each night)
5. Bell stand, which includes bellpeople and courtesy car service

Failure Identifier

Manager Hamid has been quite concerned of late, because he has been receiving numerous complaints about the Tiara Plaza courtesy car

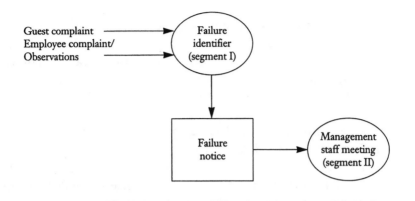

Figure E.2. Failure Identifier: Hotel Operations

service. It is the responsibility of the courtesy car service to pick up hotel guests when they arrive at the San Francisco International Airport and to shuttle them to the hotel for check-in. Also, the hotel courtesy car transports guests back to the airport after they have checked out.

The 10-mile trip from the airport can take anywhere from 30 to 60 minutes round-trip, depending on the traffic. Traffic becomes very heavy during the morning and evening rush hours.

The most common — and most serious — complaint stems from the inflexible hours of service operation. The courtesy car only leaves for the airport on the hour, which often results in the following situations detrimental to guests' perceptions of the hotel service:

1. Guest is very nervous during the trip to the airport and arrives just barely in time for the flight.
2. Guest arrives at the airport too late for the flight.
3. Guest becomes very upset and decides to call a cab to ensure that he or she gets to the airport on time for the flight.

Complaints arising due to the courtesy car scheduling most frequently are made directly to the courtesy car drivers, but occasionally are addressed in letters to hotel management.

All employees having contact with hotel guests are instructed on how to record guest complaints or their observations regarding hotel operations. Failure Notices are completed (see Figure E.2) and contain the following information:

1. Name of employee completing the failure notice
2. Date of complaint
3. Name of guest/employee with complaint/observation, if different from (1) above
4. Nature of complaint /observation

The notices (complaints) are filled out by courtesy car drivers and given to the Front Office Supervisor so that the issue may be addressed at the management staff meeting. Hamid knows the courtesy car issue will be on the agenda for the next staff meeting, so he meets with Bell Captain Bongo to get any additional pertinent information.

Bongo has been the bell captain for four years. The bellstand has recently had a big turnover in personnel. As a result, courtesy car drivers are sometimes required to help guests check in, and bellpeople are sometimes needed to make trips to the airport. Bongo apologizes profusely to his boss, as Hamid heads off to the meeting.

Management Staff Meeting (Segment II)

At the Management Staff Meeting (MSM; see Figure E.3), a time slot is always scheduled for discussion of complaints and corresponding Failure Notices. Front Office Manager Hamid initiates discussion on the complaints related to the courtesy car service. He begins by giving an introduction to the policies and procedures of courtesy car operation and then summarizes the complaints.

Other department managers participate in the ensuing discussion. The coffee shop manager, Yolanda, speaks of a guest who dashed away from a half-eaten breakfast without paying: "He seemed fine until he heard the announcement that the hotel courtesy car was leaving for the airport," Yolanda explained. "It was only then that he learned the courtesy car leaves on the hour, and *not* at the whim of the guests."

Depending upon the degree of the problem, Process Improvements and Corrective Actions may be determined at the current meeting. In this case, however, it is decided that a comprehensive analysis is needed, and a report should be prepared.

A specific statement of the problem will be required. A determination of the probable causes must be made, and several Corrective Actions and Process Improvement Recommendations should be developed.

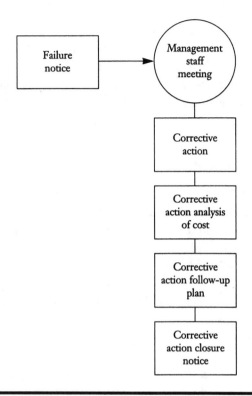

Figure E.3. Management Staff Meeting

In addition, those making the Process Improvement Recommendations are asked to prepare Corrective Action and Process Improvement Costs Analyses for each one. Individuals are assigned responsibility for these inputs, due at the next weekly meeting. Until that time, Bell Captain Bongo will be given the authority to hire a taxi if needed to satisfy guest transportation requirements. The Hotel Controller points out that the payment to the taxi driver will come out of the hotel petty cash fund, which is an overhead expense. One week later, at the next meeting, the following data were presented.

Statement of Problem

Guests frequently miss flights or must hire a taxi at their own expense. Some bellpeople even charge guests to take them to the airport in their own cars, as the courtesy car may only be driven on the hour. Management, because of insurance ramifications, forbids this action, and, when it is discovered, the bellperson is subject to disciplinary action. Guests also complain that after a long flight, they often must wait on the curb at the airport for as much an hour before they are picked up by the next scheduled courtesy car. Thus, the problem has been stated clearly.

Determination of Probable Causes

1. *Cause A:* Hotel policy is inadequate. The courtesy car is only allowed to leave on the hour. A guest arriving for the courtesy even one minute late may have to wait 59 minutes until the next departure or pick-up.
2. *Cause B:* Even if the hotel policy regarding departure to the airport is indeed adequate, it is not being communicated to guests. Often the complaint is "I could have gotten to the courtesy car departure area earlier had I known the hotel policy."
3. *Cause C:* In the case of people waiting to be picked up at the airport, the PBX operator (who receives calls for airport pick-ups) does not successfully communicate guests' needs to the courtesy car drivers. When a guest does not need a ride to the airport the courtesy car driver is unaware that someone is at the airport waiting, the courtesy car does not make the trip to the airport.

So, the problem has been stated, and potential causes have been offered.

Brainstorming

Brainstorming follows in an attempt to determine which are the most likely causes, the probable root causes, and corresponding Corrective Actions. During this discussion, the cost analyses are considered. Probable discussion points may include the following:

1. Why are many guests under the impression that the courtesy car leaves the moment a guest needs it?
2. Why aren't all guests informed that the courtesy car leaves only on the hour?
3. Has there been any benchmarking performed? Do other hotels offer courtesy car services that leave at the whim of the guests?

4. Possibly some bellpeople intentionally mislead guests so that they can drive the guests to the airport in their own cars and charge an amount that would easily exceed the usual tip.
5. Perhaps a policy of cars leaving every hour on the hour is not adequate. It is true that many hotels located within 5 miles of the airport have full-time courtesy car drivers that constantly drive back and forth and thus have a car leaving almost every 5 minutes. But, a round-trip to the airport from this hotel averages one-half to one full hour. It would require six to ten courtesy cars in operation simultaneously to match the frequency offered by these other hotels. The cost of concurrent operation of six to ten courtesy cars would be quite high. Would this be practical and cost effective?

A potential policy change could result in maintaining the hourly departure time but allow courtesy car departures for pick-ups at the airport as needed. Such a policy change would eliminate guests waiting for an hour at the airport. Perhaps the current courtesy car policy is valid and sufficient, as long as the policy is presented clearly to hotel guests.

Identification of Process Improvements

Three Process Improvements are then identified:

1. Revise policy.
2. Take steps to ensure guests are properly informed of hotel policy.
3. Allow the bell captain to hire a taxi every time a courtesy car is needed but none is available.

Discussion regarding these potential Corrective Actions focuses on two elements: (1) quality of service, and (2) cost of service. As the pros and cons of each are debated, special consideration is given to the cost impacts.

Adoption of Process Improvements

It is agreed to adopt — to some extent — all three of these Process Improvements. The following steps are designated and responsible managers assigned:

1. Courtesy car policy is revised to allow immediate pick-ups of guests at the airport. The PBX operator is directed to notify courtesy car drivers when a guest is at the airport for pick-up. This is coordinated with registration and cashiers at the front desk. If the PBX operator is unable to achieve telephone contact with a courtesy car driver, the operator may assign the responsibility of contacting a courtesy car driver to an employee at the front desk. In this case, the front desk personnel would keep an eye out for the first bellperson to return to the lobby. If no bellperson returns within 5 minutes of the call, the PBX operator will page a bellperson to come to the front desk.
2. No revision is made to the policy of making departures on the hour.
3. Bell Captain Bongo holds a meeting with all bellpeople and courtesy car drivers to ensure that they understand the policy. Attendees are also instructed to inform guests frequently of the policy. Such informing should be performed as part of the service when picking someone up at the airport or when assisting a guest checking into their room.
4. Bellpeople are instructed to include as part of the check-in service, the following:
    a. Escort guest to room.
    b. Set up their bags on luggage racks.
    c. Point out the location of ice and vending machines and offer to get ice.
    d. Inform guest of any required safety information.
    e. Announce restaurant and lounge hours of operation.
    f. Explain procedure for getting wake-up calls.
    g. Offer to set the guest up with a rental car or sightseeing tour.
    h. Clearly tell guest that *the courtesy car leaves for the airport every hour on the hour.* Suggest that the guest call down 5 to 10 minutes early so the bellperson can come up and assist with checkout and can see to it that the driver will be ready to go on the hour.
    i. Provide miscellaneous other information.
    j. Entertain questions.
    k. Hand guest the room key.
    l. Accept tip graciously.

5. Bellpeople and courtesy car drivers will be cross-trained to be able to perform either task when necessary and to understand the policies of both functions.

6. Large signs will be posted at the bellstand and at the courtesy car departure area. The signs announce that the courtesy car departs every hour on the hour.

7. The PBX operator will make announcements over the hotel public address system 15, 10, 5, and 1 minute prior to courtesy car departure for the airport.

8. Even after all these actions are taken, the bell captain (or acting bell captain) is still authorized to hire taxis for guests whenever it is felt that a guest was not adequately informed of departure times.

Cost Analyses

Lastly, cost analyses are finalized for the chosen actions. The analyses contain the following data:

1. Cost of performing the new task, such as having the courtesy car operate on a will-call basis for airport pick-ups. Costs are stated in two elements:
    a. Labor costs, stated in terms of hours and fractions thereof
    b. Nonlabor costs, itemized by expenditure

2. Costs related to the previous policy — this is the cost of having just one airport trip each hour.

To determine the cost delta, prior costs can be subtracted from the new costs, resulting in the cost impact of the Corrective Actions. The Hotel Controller Office will often work with the corresponding hotel department in preparing the cost analyses. This collaboration ensures a more accurate estimate of costs.

When more than one Corrective Action can solve a specific problem, a comparison of cost analyses will help determine the most cost-effective solutions. In this case study, the comparison was made before selecting and implementing Corrective Actions.

With a full understanding of the cost ramifications, management considers the costs involved in implementing the Process Improvement and contrasts this against what the hotel will gain (or not lose) in respect to customer satisfaction. There needs to be agreement that the costs involved will result in an equal or greater return in income.

Development of Corrective Action Follow-Up Plan

The meeting attendees generate a follow-up plan for each unique Corrective Action or Process Improvement. The follow-up plan allows the company to increase profitability in three ways:

1. If the Process Improvement/Corrective Action did not fully satisfy the cause of failure, follow-up action in a timely manner can lead to generation of an effective alternate Process Improvement.
2. If a Process Improvement has proven to be successful, follow-up action can alert staff meeting attendees that this process may be useful at other hotels or for other applications. For example, the technique of using the hotel paging system may make other helpful announcements.
3. By following up, processes that are no longer needed can be readily identified. For example, after 3 months, if only those guests requiring special airport taxi rides are those who carelessly oversleep, perhaps the on-call taxi service should no longer be allowed.

The cost analyses and the follow-up plan are provided to the Hotel Controller Office. When it is determined that a particular Process Improvement is no longer required (as in the taxi example above), and that failure has been corrected, the Closure Notice is generated. The Closure Notice officially completes the Financially Focused Quality process by modifying and/or formalizing the Process Improvements into routine procedures. The Closure Notice is provided to the Hotel Controller Office.

Hotel Controller (Segment III)

As shown in Figure E.4, the responsibilities of the controller increase in a Financially Focused Quality environment. The controller receives the cost analyses, follow-up plans, and Closure Notices from the management staff meeting. The primary additional tasks include:

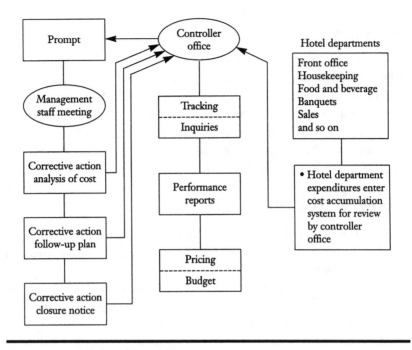

Figure E.4. Hotel Controller Office

1. Considering changes to pricing (e.g., room rates) and budgets based on the cost analyses.

2. Using analyses of cost — when pricing a new product or contract, the controller relies heavily on historical data. It is advantageous to alert the controller to the cost impacts of new Corrective Actions. The controller's office considers revising prices to ensure that any additional operating costs are included.

3. Prompting management regarding cost analyses, follow-up plans, and Closure Notices — the controller is in the ideal position to prompt management when expenditures are not in line with the cost analyses. Perhaps the performing organization has misunderstood the Corrective Action and is not performing properly. The controller, in reviewing cost data, can identify this situation and work with management toward improving the cost effectiveness of the Corrective Actions.

4. Finally, ensuring that management is adhering to the follow-up plan by issuing alternative Corrective Actions or Closure Notices in a timely manner.

Summary

The key to a successful cost-effective resolution of the courtesy car failure is not found in any one Financially Focused Quality tool. It does not matter whether the problem situation is reported on a Failure Notice or a Corrective Action Recommendation form. The important factor is the overall Financially Focused Quality mindset. Every member of Tiara Plaza management is required to examine the cost impacts of Process Improvements and allow such considerations to heavily affect the final selection process. Financially Focused Quality, as outlined here, presents the Tiara Plaza with the means to:

1. Determine the existence of specific problems.
2. Take timely and appropriate Corrective Action to ensure such problems are resolved.
3. Ensure the most economical implementation of Corrective Actions through cost comparisons, Corrective Action revisions, and cancellations/formalizations.
4. Anticipate increased operating costs in preparation for salary and wage negotiations, and revise pricing policies.

It has been proven that effective use of FFQ tools can enable companies to increase productivity and profitability. This case study exhibits just one of the many potential applications of the Financially Focused Quality approach to Process Improvement and Corrective Action.

Self-Study/Discussion Questions

1. Would you recommend initiating formal financial training for employees in hotel operations? Why or why not? Which employees, if any, would receive such training, and by what method?
2. Describe the FFQ activity that might take place involving complaints about broken icemakers or vending machines. Identifiers? Potential and probable causes? Potential Corrective Actions? Cost analyses? Implementation? Follow-up and closure?
3. Describe FFQ activity for complaints about coffee shop employees, about the time it takes to prepare and serve dinner in the dining room, or lost reservations.
4. What other problems in hotel operations have you experienced? How could they be resolved using FFQ?

Index

A

Accelerated test, 79
Accounting classifications, 142
Accounting
    cost, 175
    financial, 175
    general, 173
    payroll, 173–174
Accounts payable, 174–175
Accredited laboratories, 129–130
Audit checklist, for basis of estimate,
    169–170
Audit control, 130
Affinity Diagrams, 2, 28

B

Basis of estimate, 166–168
    audit checklist for, 169–170
Benchmarking, 2, 31, 39–40
Benefits. *See* Employee
Best practices, 2, 31, 39
Beta videotape format, 51
Bill of Materials (BOM), 35
Bob's Purple Bayou Café, 257–264
BOM. *See* Bill of materials
Brainstorming, 27, 216–222
    components of, 221
    presentation of rules in, 218
    recording ideas in, 218
    topic or objective definition in, 217

Budget cutting, 32, 44–45
Budgeting, 170–171
Business objectives, 31, 36
    of Bob's Purple Bayou Café, 258
Business process improvement, 3

C

Cadbury Schweppes, 5
CAIV. *See* Cost as an Independent
    Variable
California State University, Monterey
    Bay, Vision Statement of, 37–39
Centers of Excellence, 265
Charge numbers, 145
CI. *See* Continuous Improvement
Citicorp, 5
CLCA. *See* Closed-Loop Corrective
    Action
Closed-Loop Corrective Action (CLCA),
    68–69
    fleet and field facilities, 74
    operations and failure reporting
        systems, 70
    quality engineering in, 74
    relationship to Financially Focused
        Quality, 70
    supplier operations, 70
Commercial contracting, 163–164
Communications, 3
Commute alternatives, 13

Computer
   Macintosh vs. Personal, 52
   products, 51
Configuration management, 79
Conforming to requirements, 102–103
Consumerism, 126
Continuous Improvement, 2
Continuous Quality Improvement
   (CQI), 3, 18–19
Contract closure, 176
Contracting mode, 154
Core competencies, 31, 42
Corporate world, 5
Corrective Action Follow-Up Plan,
   224–226. *See also* Process
   Improvement Follow-Up Plan
   form example, 225
   in hotel operations, 284
Cost accumulation structure, 145
Cost analysis, in hotel process
   improvement, 283
Cost as an Independent Variable
   (CAIV), 32, 46
Cost-effective quantities, ordering in, 102
Cost modeling, 269–270
Cost Performance Integrated Product
   Team (CPIPT), 46
Cost plus award fee, 156
Cost plus fixed fee, 155
Cost plus incentive fee, 155
Cost reduction/avoidance programs, 3, 18
Cost reimbursable contract, 155
Cost segregation, 146
Costs, direct-direct and allocated-direct,
   144
   direct vs. indirect, 142
   labor vs. nonlabor, 143
   overhead, 144
   poor quality, 32, 45–46
   quality, 3, 32, 45–46
CPIPT. *See* Cost Performance Integrated
   Product Team
CQI. *See* Continuous Quality
   Improvement
Customer
   and inexpensive/expensive products,
      56–57
   feedback, 56–57
   focus, 2, 49–62
   perception, 2
   satisfaction, 2
Cycle of financial activities, 139–141
Cycle of quality activities, 77
   finance in, 84–85

D

Degree of inspection, 124
Deming, W. Edwards, 63–65
Dental benefits, 11
Department improvement team, 21–22
Design
   documentation reviews, 98–99
   qualification test, 100
   quality assurance, 76
Destructive testing, 80
Diminishing returns
   point of, 32
   quality control measures, 65–66
Direction
   direct vs. indirect costs, 142
   in FFQ Blueprint, 33
Disallowances, 156
Disposal phase and quality
   assurance, 77
Downsizing, 3, 32, 44–45

E

Eastman-Kodak, 5
Education, assistance with further, 12
Effective meetings, 2, 22–27
Electrical in-process inspection, 80
Employee
   benefits, 10
   participation in quality
      management activities, 3,
      14–28
   recreation facilities, 13
   stock options, 12
   stock ownership plan, 3, 11
   suggestion systems, 3, 17
Empowerment, 3, 16
Engineering evaluation test, 80
Enterprise, impacts in FFQ
   Blueprint, 35

Environment, and empowerment/
   ownership, 16
Equipment, and empowerment/
   ownership, 16
Errors, in inspection, 125
ESOP. *See* Employee stock ownership plan
Euro, 50
Executive perks, 14

F

Fabrication and assembly inspection and
   test, 80
Failure, during manufacturing, 190
   customer returns, 191–192
   hotel industry, 192
   generic, 192–193
Failure analysis, 80
Failure analyst, 181–182, 212–214
Failure diagnosis, 80
Failure identification. *See*
   Opportunity identification
Failure Identifier, 181, 183–185
   in hotel operations, 276–278
Failure mode, in Perky Pets, 237
Failure Notice, 181, 190–191
   coordination of, 205–206, 212
   form example, 191
   in auto or aircraft manufacturing
      failure, 207–208
   in Perky Pets case study, 239
   in service industry, 208
   in television manufacturing, 208
   time requirements for
      coordination of, 205
Failure verification, 81
FCCA. *See* Full-Cycle Corrective Action
Feedback, 56–60
   at Bob's Purple Bayou Café, 263
FFQ. *See* Financially Focused Quality
Field results, 127–128
Final inspection, 79
Finance, in FFQ, 182
Financial activities, cycle of, 139
Financial administration, 137–161
   accounting classifications, 142
   budgeting, 170
   contract closure in, 176

cycle of financial activities in, 139
Financially Focused Quality and,
      182, 227–229
forecasting personnel, 150
general accounting, 173
government contracting mode, 154
Perky Pets and, 241–243
probabilities in, 220
proposals and pricing, 163–170
rates per direct-labor hour, 156
reporting in, 172
timekeeping and cost segregation,
      146–148
Financial concepts, 141
Financial functions, 163–177
Financial training, 138–139
Financially Focused Quality (FFQ), 6,
      179–231
   Blueprint, 31, 32–35
   decision-making in, 203–231
   finance and, 182, 227–229
   improvement coordination in,
      203–231
   introduction to, 137
   overview, 180
   probabilities, 220
   product support, 183, 229–230
   relationship of Closed-Loop
      Corrective Action to, 70
   software engineering in, 245–255
   training, 137–161
Financially Focused Quality mindset, 9,
      33, 206–207
Firm fixed price, 154
First article compatibility test, 81
First article inspection, 108
Fixed price contracting mode, 154–155
Fixed price incentive, 155
Flexibility with work shifts, 13
Focus
   on external customer, 55
   on quality, 2
Forecasting
   criteria, 150
   direct personnel, 151
   facilities, 150
   personnel, 150
   problems in, 154

proposals, 150
    training, 150
Forward pricing rates, 151
Fruit of the Loom, 5
Full-Cycle Corrective Action (FCCA), 3,
    32, 46
Funding fence, 171

G

General and administrative costs, 151
Government contracting, 164–165

H

Harrington, H. James, 4, 32
Historical basis of estimate, 168
Holiday benefits, 11
Hotel operations, 273–286
Human Resources, 10
    and forecasting, 151

I

Industrial quality assurance, 116
In-process inspection, 79, 118–119
    and testing, 118–119
Inspection
    degree of, 124
    errors, 125
    final, 79
    first-article, 81, 108
    in-process, 79, 80, 118
    instructions, 114–115
    mechanical in-process, 81
    obligation for, 111
    100% receiving, 109–110
    procedure manuals, 115
    records during the receiving cycle, 110
    shipping, 79, 126
    source, 108
    storage facilities, 110–111
    test areas, 111–112
    vendor, 109
    vendor components, 107
Inspectors
    errors of, 125–126
    training of, 113

Interrelationship diagrams, 2
ISO series, 52–55
    9000, 2, 52–54
    14000, 2, 52–55

J

Just-in-Time (JIT), 3, 31
    manufacturing, 36

K

Kaizen events, 2, 28

L

Labor recording, 148
Lean process, 3, 28
Letters, orchid/unsolicited, 60
Levi-Strauss, 5
Little, Arthur D., 4
Loss leaders, 10

M

Machine capacity planning, 36
Machining Outsourcing Initiative
    (MOI), 266
Macintosh computer, 52
Make-or-buy committees, 101
Malcolm Baldrige National Quality
    Award criteria, 2, 4
Management reserve, 171–172
Management resources in assuring
    quality, 129
Managers as team builders, 3
Manual timecards, 148
Manufactured product quality, 93
Manufacturer liability, 88
Manufacturing failure, 207–208
Manufacturing process, 117
Manufacturing Resource Planning
    (MRP2), 2, 31, 36
Material requirements planning, 2
Material Review Board, 75, 81
Matrix diagrams, 2
Measurement process, 31, 41–42, 112–113
Measurements, 35

Mechanical in-process inspection, 81
Medical and dental benefits, 11
Meetings
    at the right time and place, 25
    effective, 22
    locations of, 26
    roles of attendees, 26
Metrology audits, 81
Microsoft Office, 41
Mission Statement, 3, 31, 39
    Bob's Purple Bayou Café, 259
MOI, *See* Machining Outsourcing
    Initiative
Movie formats, 51–52
MRP2. *See* Manufacturing Resource
    Planning

N

Natural Work Teams, 21–22
Newsweek, 4, 15
Nonconforming material, 119–120

O

Operation phase, and quality assurance,
    77
Opportunity identification
    customers and, 188
    government representatives and, 188
    in the factory, 189
    methods for, 185
    process, 186
    warranty service organization and,
        189
Opportunity Identifier, 181, 183–185
    in hotel operations, 273–274
Organizational structure, 35
Outsourcing, 3, 31, 42–43, 265–272
Overhead (indirect) costs, 144
Ownership, 3, 15, 16

P

Packaging for product safety, 126–127
Participative management, 20
Partnerships with suppliers, 32, 45
Parts application review, 81

Pay for performance, 3, 19–20
Payroll accounting, 173–174
Penguins-in-a-Helmet, 233–244
Perks, executive, 14
Perky Pets, 233–244
    Financially Focused Quality at,
        237–244
    manufacturing process of, 234
    marketing of, 235
Personnel forecasting, 150
PIC. *See* Process Improvement
    Coordinator
PICA. *See* Process Improvement Cost
    Analysis
PICN. *See* Process Improvement Closure
    Notice
PIFP. *See* Process Improvement Follow-
    Up Plan
PIR. *See* Process Improvement
    Recommendation
PM. *See* Process Management
Point of diminishing returns, 32
Poor quality cost, 3
Pricing, 160–170
Prioritization matrix, 2
Probabilities, in FFQ, 220–221
Process Analysis Meeting, 182, 214–216
Process analyst, 181–182
Process breakthrough, 34
Process control, 117
Process decision program charts, 2
Process evaluation, 116–117
    mandatory process evaluation, 116
Process Improvement, 3, 182
    when not required, 204
Process Improvement closure, 183
    in software engineering, 25
Process Improvement Closure Notice
    (PICN), 226–227
    form example, 227
Process Improvement coordination,
    203–206
    auto or aircraft manufacturing
        failure, 207
    machining outsourcing initiative, 268
    service industry, 208
    software engineering, 252–253
    television manufacturing, 208

time requirements for, 205
Process Improvement Coordinator
    (PIC), 181
    related functions of, 206
Process Improvement Cost Analysis
    (PICA), 182, 222–224
    form example, 223
Process Improvement Follow-Up Plan
    (PIFP), 182, 224–226
    form example, 225
Process Improvement
    Recommendation (PIR), 181,
    193–202
    coordination of PIR concerns,
        210–211
    coordination of PIR projects, 209
    ease of processing, 195
    examples of, 200
    form example, 193
    illustration of acting on, 201
    low implementation/
        administration cost, 195
    mandatory management
        involvement, 196
    motivation, 198
    related functions of the PIC, 206
    success factors of, 195
Process Improvement Team, 22
Process Management (PM), 3, 19
Process quality control, 65
Product design, 96–99
        safety, 87, 93, 96
Product evaluation test, 81
Product processes, 34
Product recall, 128–129
Product support, in FFQ, 183, 227,
    229–230
    Perky Pets, 243
Production phase, and quality assurance,
    77
Production process
    assessment of, 116
    capability studies, 116
    mandatory evaluation, 116
Product liability, 87–134
    basic conditions of, 89
    design and development, 93
    manufacture and use, 107–134

manufacturer liability, 88
manufacturing defects, 90
relationship to quality assurance, 90, 92
Product life-cycle, quality assurance and,
    76–77
Product safety, 87
    after delivery, 94
    audit, 131–131
    classification of critical
        characteristics, 100
    design review, 93, 99
    evaluation of product design, 98
    Financially Focused Quality
        approach to corrective action,
        96
    in the field, 127
    packaging for, 126
    related quality information, 95
    responsibility for, 88
    specifications, 97
Profit generation, in FFQ Blueprint, 34
Proposal components, 166
Proposal preparation flow, 165–166
Proposals and pricing, 163–170
    commercial contracting, 163
    engineering estimates, 168
    government contracting, 164
    historical basis of estimate, 168
    security guidelines, 168
Purchased material quality, 93
Purchasing provisions and
    specifications, 104–105

Q

Quality
    becoming big business, 3
    business of, 4
    cost of, 3, 45
    focus on, 2
    poor quality costs, 3, 45
Quality activities, cycle of, 77
Quality assessment, 82
Quality assurance
    evaluation checklist, 132–133
    evolution of, 91
    functions in, 78
    organization, 67–68

product life-cycle, and, 76
relationship to product liability, 92
statistics and sampling, 120–125
systematic approach to, 92
vs. product liability, 90
Quality audit, 82
Quality circles, 3, 20–21
Quality control, 63–86
definition and goals, 67
diminishing returns of, 65
product liability aversion, 95
Quality costs. *See* Costs
Quality management systems, 2,
49–62
Questionnaires, 58

R

Random numbers, 121
Rates per direct labor hour, 156–157
Rath and Strong, 4
Raw resource, 166
Receiving inspection
corrective action, 110
quality control, 78
test, 82
Recreation facilities, 13
Reengineering, 3, 32, 43
Reporting, 172–173
Research and development, and quality
assurance, 76
Restructuring, 3, 32, 43–44
Retirement plan, 10
Rewards and recognition, in FFQ
Blueprint, 35
Rightsizing, 3, 32, 44

S

Safety. *See* Product safety
Salary continuation, 11
Sampling inspection, 121
Sampling plans, 123–125
arguments against, 125
vs. 100% inspection, 124
Security guidelines for cost
proposals, 168–169
Self-paced training, 160

Service industry feedback, 60
Service process, 34
Service publications, 127
Shipping inspection, 79, 126
Sick pay, 11
Silicon Graphics, 5
Single 8, 51
Six Sigma, 3, 29
Smartsizing, 44
Software engineering, FFQ in, 245–255
Source acceptance, 82
Source inspection, 108
Source/receiving inspection, and
corrective action, 110
Source verification inspection, 82
SPC. *See* Statistical process control
Standardization, 50
Standards for design and
manufacture, 97
Statistical analysis, 65
Statistical process control (SPC), 82
Statistics, and quality assurance,
120–125
Stock options, 12
Stock ownership plan, 3, 11
Suggestion systems, 3, 17
Super 8 film, 51
Suppliers
evaluation questionnaire, 104
partnerships with, 45
performance history of, 103
quality evaluation survey of, 103–104
selection of, 102
Surveys, in service industry, 58–60

T

Task force teams, 21
Teaching. *See* Training
Team Building, 2, 20
Teams, types of
Natural Work, 21
Process Improvement, 22
Quality Circles, 20
Task Force, 21
Timecards, 146–148
manual vs. virtual, 148
Timekeeping, 146–148

Tolerances, 119
Tom and Bob, 58–60
Total Improvement Management, 2, 31, 32
    pyramid, 32
Total Quality Management (TQM), 2, 15
Total Quality Service, 15–16
TQM. *See* Total Quality Management
Traditional finance, involvement in cycle
    of quality activities, 83
Training
    Bob's Purple Bayou Café, 260
    empowerment and ownership, 16
    examples of FFQ materials for, 160
    financial, 138–139
    Financially Focused Quality, 137–161
    Financially Focused Quality training
        guideline, 158–160
    forecasted personnel, 150
    inspectors, 113–114
    programs, 13, 31, 40–41
Transportation alternatives, 13
Tree diagrams, 2

U

Undistributed budget, 171–172
Unsolicited letters, 60

V

Vacation benefits, 11
Value engineering, 82
Vendor
    audit, 130–131
    components, inspection and test of,
        107
    inspection, 109
    materials control, 101–104
VHS videotape format, 51
Video, 51–52
Virtual timecards, 148
Vision Statement, 3, 31, 36
    Bob's Purple Bayou Café, 258
    California State University, Monterey
        Bay, 37–39

W

Work breakdown structure, 144–145
Work shifts, flexibility of, 13

Y

Yield/realization factor, 152–154